Lecture Notes in Mathematics

Edited by A. Dold, Heidelberg and B. Eckmann, Zürich

T0233144

397

Toshihiko Yamada

Tokyo Metropolitan University, Fukazawa-Cho Setagaya, Tokyo/Japan

The Schur Subgroup of the Brauer Group

Springer-Verlag
Berlin · Heidelberg · New York 1974

AMS Subject Classifications (1970): Primary: 20C05, 20C15
Secondary: 12A99, 12B99

ISBN 3-540-06806-6 Springer-Verlag Berlin · Heidelberg · New York
ISBN 0-387-06806-6 Springer-Verlag New York · Heidelberg · Berlin

Offsetdruck: Julius Beltz, Hemsbach/Bergstr.

PREFACE

These notes are taken from a course on theory of the Schur subgroup of the Brauer group which I gave at Queen's University in 1971/72, including subsequent developments of it.

The study of the Schur subgroup or Schur algebra was begun by I. Schur at the beginning of this century. But it was not until 1945 that the long surmised conjecture: "Every irreducible representation U of a finite group G of order n can be written in the field of the n-th roots of unity," was solved by R.Brauer. Early in 1950's, R. Brauer and E. Witt, independently, found that questions on the Schur subgroup are reduced to a treatment for a cyclotomic algebra. The result has been called the Brauer-Witt theorem, and we can now say that almost all detailed results about Schur subgroups depend on it.

After the Brauer-Witt theorem appeared, there was little progress until the end of 1960's. However, in a couple of years, the Schur subgroup has been extensively studied by many people, and it seems that the development has reached a certain culminating point with the most recent results. Here we will mention some of them: The Schur subgroup was completely determined for an arbitrary local field; a simple formula for the index of a p-adic cyclotomic algebra was obtained; the Schur subgroups are determined for several cyclotomic extensions of the rational field Q; some remarkable properties of a Schur algebra were discovered; etc. Thus it seems timely to undertake the task of clarifying the theory of Schur subgroup systematically, focussing on the recent progress.

I would like to thank Professor P. Ribenboim who let me give a course on this subject at Queen's University and suggested editing these notes for publication.

I am also grateful to McGill University, Queen's University, Oregon State University, and the National Science Foundation, for financial assistance at various stages of this work.

TABLE OF CONTENTS

INTRODUCTION

Let k be a field of characteristic 0. Denote by $Br(k)$ the Brauer group of k. Let A be a central simple algebra over k. Then $[A]$ is the algebra class of $Br(k)$ which contains A. Denote by A^* the multiplicative group of invertible elements of A. If there exists a finite subgroup G of A^* such that A is spanned by G with coefficients in k, i.e., $A = \{\sum_{g \in G} a_g g;\ a_g \in k\}$, then A is called a Schur algebra over k. It is easy to see that A is a Schur algebra over k if and only if A is a simple component (central over k) of the group algebra $k[H]$ for some finite group H. The Schur subgroup $S(k)$ of the Brauer group $Br(k)$ consists of those algebra classes that contain a Schur algebra over k. We face two questions: (i) Given a field k, determine the Schur subgroup $S(k)$. (ii) Find properties of a Schur algebra over k. We will see how the Brauer-Witt theorem is fundamental for solving these questions.

Consider the following crossed product B over k:

$$B = (\beta,\ k(\zeta)/k) = \sum_{\sigma \in G} k(\zeta)u_\sigma \qquad \text{(direct sum)}$$

$$u_\sigma u_\tau = \beta(\sigma,\tau)u_{\sigma\tau}, \quad u_\sigma x = x^\sigma u_\sigma \quad (x \in k(\zeta)),$$

where ζ is a root of unity, $G = G(k(\zeta)/k)$, and β is a factor set (2-cocycle) of $k(\zeta)/k$ such that for every σ, τ of G, $\beta(\sigma, \tau)$ is a root of unity contained in $k(\zeta)$. Such a crossed product is called a cyclotomic algebra (Kreisalgebra) over k. Let C(k) denote the set of all those algebra classes of Br(k) which are represented by a cyclotomic algebra over k. Then C(k) is a subgroup of Br(k). Thus, we have two subgroups, S(k) and C(k), of Br(k). But it follows from the Brauer-Witt theorem that S(k) = C(k). Consequently, we have only to study cyclotomic algebras over k on all matters pertaining to the Schur subgroup S(k).

Let p be a prime, Q_p the rational p-adic numbers, and k a cyclotomic extension of Q_p. Let B be a cyclotomic algebra over k. By detailed number-theoretical calculation of 2-cocycles, the author obtained a simple formual for the index of B. By making use of it, the Schur subgroup S(k) was completely determined. It turns out that if $Q_p(\zeta)$ is a cyclotomic field containing k, then the structure of S(k) is given in terms of the inertia group of the extension $Q_p(\zeta)/k$, provided that $Q_p(\zeta)$ contains a primitive p-th (respectively, 4-th) root of unity for $p \neq 2$ (respectively, for $p = 2$). It can be said that one of the purposes of this set of notes is to give a full account of the theory of Schur subgroups for local fields, which is, in addition to its own interest, indispensable for detailed study on Schur subgroups of global fields.

Here we will briefly explain the contents of these notes. Chapter 1 contains basic facts about Schur algebras, which are not readily found in the literature. In Chapter 2, cyclotomic algebras are studied. It is shown that each cyclotomic algebra is associated with a factor set which is fit for computation. Furthermore, the defining relations of a cyclotomic algebra are stated with respect to this factor set. A complete account of the Brauer-Witt theorem is given in Chapter 3. A formula for the index of a p-adic cyclotomic algebra, as well as the determimination of the Schur subgroup of a local field, is given in Chapters 4 and 5. Several properties of a Schur algebra are stated in Chapter 6. The study of Schur subgroups for algebraic number fields is done in Chapters 7 and 8. Chapter 9 contains qualitative theorems about splitting fields of group algebras and the relation between the Schur index and the exponent of a group, etc.

Chapter 1. SCHUR ALGEBRAS

Throughout this chapter, k is a field of characteristic 0 and $\bar{k}$ its algebraic closure. Let G be a finite group and χ an irreducible character of G. Then the simple component $A(\chi, \bar{k})$ of $\bar{k}[G]$ which corresponds to χ is:

$$A(\chi, \bar{k}) = \bar{k}[G]e(\chi); \quad e(\chi) = \chi(1)|G|^{-1} \sum_{g\in G} \chi(g^{-1})g, \tag{1.1}$$

$e(\chi)$ being a primitive central idempotent of $\bar{k}[G]$. (See [15, §33].) By orthogonality relations for characters, we have

$$\chi(e(\chi)) = \chi(1)|G|^{-1} \sum_{g\in G} \chi(g^{-1})\chi(g) = \chi(1),$$

$$\chi(e(\psi)) = \psi(1)|G|^{-1} \sum_{g\in G} \psi(g^{-1})\chi(g) = 0$$

for any irreducible character $\psi \neq \chi$ of G. Note that $e(\chi) \in k(\chi)[G] \subset \bar{k}[G]$. Let $\sigma \in \mathcal{G}(\bar{k}/k)$. Then σ induces the automorphism of $\bar{k}[G]$ over k, which is also denoted by σ:

$$\sigma: \sum_{g\in G} \alpha_g g \to \sum_{g\in G} \alpha_g^\sigma g, \quad (\alpha_g \in \bar{k}).$$

We notice that $\sigma(e(\chi)) = e(\chi^\sigma)$.

Proposition 1.1. Notation being as above, the simple component $A(\chi, k)$ of $k[G]$ corresponding to χ is:

$$A(\chi, k) = k[G]a(\chi); \quad a(\chi) = \sum_{\tau\in\mathcal{G}} e(\chi^\tau), \quad \mathcal{G} = \mathcal{G}(k(\chi)/k).$$

$a(\chi)$ is a primitive central idempotent of $k[G]$.

Proof. We see that

$$a(\chi) = \sum_{\tau \in G} e(\chi^{\tau}) = \chi(1)|G|^{-1} \sum_{g \in G} \left(\sum_{\tau \in G} \chi^{\tau}(g^{-1})\right) g \in k[G],$$

and that $k[G] \subset \bar{k}[G]$. So, $a(\chi)$ is a central idempotent of $k[G]$. If c is a central idempotent of $k[G]$, then c is also a central idempotent of $\bar{k}[G]$. The primitive central idempotents of $\bar{k}[G]$ are precisely the $e(\psi)$, where ψ ranges over all the irreducible characters of G. Each central idempotent c' of $k[G]$ is of the form: $c' = e(\psi_1) + \cdots + e(\psi_s)$, where $\psi_1, \cdots, \psi_s$ are distinct irreducible characters of G, uniquely determined by c'. It follows immediately that if $a(\chi)$ is not primitive as a central idempotent of $k[G]$, then there exist non-empty subsets I and J of the set $\mathcal{G} = \mathcal{G}(k(\chi)/k)$ such that $\mathcal{G} = I \cup J$, $I \cap J = \phi$, and that

$$a(\chi) = a_1 + a_2, \quad a_1 = \sum_{\tau \in I} e(\chi^{\tau}), \quad a_2 = \sum_{\tau' \in J} e(\chi^{\tau'}),$$

$a_1, a_2 \in k[G]$. Say $I \ni 1$, $J \ni \sigma$. Then, $a_1 = a_1^{\sigma} = \sum_{\tau \in I} e(\chi^{\tau\sigma})$, and so $e(\chi^{\sigma})$ appears in the summands of a_1, contradiction. Thus, $a(\chi)$ is a primitive central idempotent of $k[G]$. We have

$$\chi(a(\chi)) = \sum_{\tau \in \mathcal{G}} \chi(e(\chi^{\tau})) = \chi(e(\chi)) = \chi(1) \neq 0,$$

and so $\chi(k[G]a(\chi)) \neq 0$. #

Corollary 1.2. If $k(\chi) = k$, then $A(\chi, k) = k[G]e(\chi)$.

Proof. This is obvious, because if $k(\chi) = k$, then $e(\chi) = a(\chi)$. #

Let U be an irreducible representation of G of degree n and χ the character of U. We assume that U is realized in $\bar{k}$. Then U can be regarded as a representation of $k[G]$. The enveloping algebra $\mathrm{env}_k(U)$ of U over k is defined as the image of $k[G]$ by U:

$$\mathrm{env}_k(U) = U(k[G]) = \{ \sum_{g\in G} \alpha_g U(g);\ \ \alpha_g \in k\} \subset M_n(\bar{k}).$$

It is wellknown that if $k = \bar{k}$, then $\mathrm{env}_{\bar{k}}(U) = M_n(\bar{k})$.

Proposition 1.3. Notation being as above, $k[G]a(\chi)$ is k-isomorphic to $\mathrm{env}_k(U)$.

Proof. We know that $U(e(\chi)) = 1_n$, the identity matrix of $M_n(\bar{k})$, and that $U(e(\psi)) = 0$ if ψ is an irreducible character of G different from χ. Hence, $U(a(\chi)) = \sum_{\tau\in G} U(e(\chi^\tau)) = U(e(\chi)) = 1_n$, and so $U(k[G]a(\chi)) = U(k[G])\cdot U(a(\chi)) = \mathrm{env}_k(U)$. Thus U induces a k-algebra homomorphism of $k[G]a(\chi)$ onto $\mathrm{env}_k(U)$. This must be an isomorphism, because $k[G]a(\chi)$ is simple. #

For a ring A, $Z(A)$ denotes the center of A. For $g \in G$, $C(g)$ denotes the conjugate class of G containing g.

Proposition 1.4. The center of $k[G]a(\chi)$ is k-isomorphic to $k(\chi)$. The isomorphism is given by

$$\chi(1)|C(g)|^{-1}(\sum_{h\in C(g)} h)a(\chi) \rightarrow \chi(g).$$

The center of $env_k(U)$ is $k(\chi)\cdot 1_n$. If $k = k(\chi)$, then the center of $k[G]e(\chi)$ is $k\cdot e(\chi) \cong k$.

Proof. If $\omega \in Z(k[G]a(\chi))$, then $\omega \in Z(k[G]) \subset Z(\bar{k}[G])$. Hence $U(\omega) \in Z(U(\bar{k}[G])) = Z(M_n(\bar{k})) = \bar{k}\cdot 1_n$. So, $U(\omega) = \alpha\cdot 1_n$ for some $\alpha \in \bar{k}$. Taking traces of both sides of this equation, we have $\chi(\omega) = n\alpha$, and so $\alpha = \chi(\omega)/n \in k(\chi)$. Conversely, for each $g \in G$, set

$$\omega_g = \chi(1)|C(g)|^{-1}(\sum_{h\in C(g)} h)a(\chi) \in k[G]a(\chi).$$

Since both $a(\chi)$ and $\sum_{h\in C(g)} h$ are central in $\bar{k}[G]$, ω_g is central in $\bar{k}[G]$ and $U(\omega_g) = \chi(1)|C(g)|^{-1}(\sum_{h\in C(g)} U(h)) = \alpha_g\cdot 1_n$ for some $\alpha_g \in \bar{k}$. Taking traces, we have $\chi(1)|C(g)|^{-1}|C(g)|\chi(g) = \chi(1)\alpha_g$, whence $\alpha_g = \chi(g)$. Thus, $\chi(g)\cdot 1_n = U(\omega_g)$. From these facts, if follows that $U(Z(k[G]a(\chi))) = Z(env_k(U)) = k(\chi)\cdot 1_n \cong k(\chi)$. This proves that $Z(K[G]a(\chi))$ is the subring $k(\omega_g;\, g\in G)$ of $k[G]a(\chi)$, which is generated by the ω_g $(g \in G)$ over k and is k-isomorphic to $k(\chi)$. If $k = k(\chi)$, then

$$U(\chi(g)e(\chi)) = \chi(g)\cdot 1_n = U(\chi(1)|C(g)|^{-1}(\sum_{h\in C(g)} h)e(\chi)), \ (g \in G).$$

Since U is a k-isomorphism of $k[G]e(\chi)$ onto $env_k(U)$, it follows that

$$\chi(g)e(\chi) = \chi(1)|C(g)|^{-1}(\sum_{h\in C(g)} h)e(\chi) \quad \text{for all} \quad g \in G.$$

Thus, $Z(k[G]e(\chi)) = k(\omega_g;\ g\in G) = k\cdot e(\chi)$. #

<u>Proposition 1.5</u>. $k[G]a(\chi)$ is k-isomorphic to $k(\chi)[G]e(\chi)$.

<u>Proof</u>. Since by Proposition 1.4, the center of $env_k(U)$ is $k(\chi)\cdot 1_n$, it follows that $env_k(U) = env_{k(\chi)}(U)$. Then, we have

$$k[G]a(\chi) \underset{k}{\simeq} env_k(U) = env_{k(\chi)}(U) \underset{k(\chi)}{\simeq} k(\chi)[G]e(\chi). \qquad \#$$

Set $\bar{U}(g) = {}^t(U(g^{-1}))$, $g \in G$. Then $\bar{U}$ is an irreducible representation of G with the character $\bar{\chi}$, defined by $\bar{\chi}(g) = \chi(g^{-1})$, $g \in G$. Note that $k(\chi) = k(\bar{\chi})$. We can define a k-linear mapping Ψ from $env_k(U)$ onto $env_k(\bar{U})$ by

$$\Psi(\sum_{g\in G} a_g U(g)) = {}^t(\sum_{g\in G} a_g U(g)) = \sum_{g\in G} a_g \bar{U}(g^{-1}), \quad (a_g \in k). \quad (1.2)$$

It follows readily that Ψ is injective, satisfying the relation $\Psi(\xi\xi') = \Psi(\xi')\Psi(\xi)$ for all $\xi, \xi' \in env_k(U)$. Note that both $env_k(U)$ and $env_k(\bar{U})$ have center $k(\chi)\cdot 1_n$, and that $\Psi(\alpha\cdot 1_n)$

$= \alpha \cdot 1_n$. Summarizing, we have

Proposition 1.6. Ψ is an anti-isomorphism of $\mathrm{env}_k(U)$ onto $\mathrm{env}_k(\bar{U})$ over their center $k(\chi)\cdot 1_n$.

The group algebra $k[G]$ has the following involution $*$ over k:

$$*: \sum_{g\in G} a_g g \to \sum_{g\in G} a_g g^{-1}, \quad (a_g \in k). \tag{1.3}$$

It is easy to see that $*(a(\chi)) = a(\bar{\chi})$ and $*$ induces an anti-isomorphism of $k[G]a(\chi)$ onto $k[G]a(\bar{\chi})$, and that the following diagram is commutative.

$$\begin{array}{ccc} k[G]a(\chi) & \xrightarrow{\ *\ } & k[G]a(\bar{\chi}) \\ U \downarrow & & \downarrow \bar{U} \\ \mathrm{env}_k(U) & \xrightarrow{\ \Psi\ } & \mathrm{env}_k(\bar{U}) \end{array}$$

If $k = k(\chi)$, then we can identify $Z(k[G]e(\chi)) = k\cdot e(\chi)$ and $Z(k[G]e(\bar{\chi})) = k\cdot e(\bar{\chi})$ with the field k. Thus we have

Proposition 1.7. If $k(\chi) = k$, then $k[G]e(\chi)$ is anti-isomorphic to $k[G]e(\bar{\chi})$ over k.

Corollary 1.8 (The Brauer-Speiser theorem). Let χ be a real-valued irreducible character of G. Then the Schur index $m_Q(\chi)$ over Q is 1 or 2. If the degree of χ is odd, then $m_Q(\chi) = 1$.

Proof. Let ι denote the automorphism of the complex numbers C over the real numbers R defined by $\iota(\sqrt{-1}) = -\sqrt{-1}$. Then, $\bar{\chi}(g) = \iota(\chi(g)) = \chi(g)$ for all $g \in G$, and so, $\bar{\chi} = \chi$. Put $k = Q(\chi)$. By Proposition 1.7, we have

$$[k[G]e(\chi)]^2 = [k[G]e(\chi)]\cdot[k[G]e(\bar{\chi})] = 1 \text{ in } Br(k). \qquad (1.4)$$

Since k is an algebraic number field, the index and exponent of $k[G]e(\chi)$ are the same, which is at most 2 by (1.4). Recall that $m_Q(\chi)$ is the index of the simple component $A(\chi, Q)$ of $Q[G]$, which is Q-isomorphic to $A(\chi, k) = k[G]e(\chi)$ (cf. Proposition 1.5). The last assertion of the corollary follows from the fact that $m_Q(\chi)$ divides $\chi(1)$. #

Proposition 1.9. Let G and H be groups and let χ and ψ be irreducible characters of G and H, respectively, such that $k(\chi) = k(\psi) = k$. Then,

$$k[G]e(\chi) \otimes_k k[H]e(\psi) \underset{k}{\simeq} k[G \times H]e(\chi \otimes \psi).$$

Proof. We have the isomorphism:

$$\Phi: \; k[G] \otimes_k k[H] \simeq k[G \times H],$$

$$\Phi((\sum_g a_g g) \otimes (\sum_h b_h h)) = \sum_{g,h} a_g b_h (g \times h), \quad (a_g, b_h \in k).$$

It is easy to see that $\Phi(e(\chi) \otimes e(\psi)) = e(\chi \otimes \psi)$, because

$$e(\chi \otimes \psi) = \chi(1)\psi(1)|G|^{-1}|H|^{-1} \sum_{g,h} \chi(g^{-1})\psi(h^{-1})(g \times h).$$

Hence Φ induces a k-isomorphism of $k[G]e(\chi) \otimes_k k[H]e(\psi) = (k[G] \otimes k[H])(e(\chi) \otimes e(\psi))$ onto $k[G \times H]e(\chi \otimes \psi)$. #

It follows from Propositions 1.7 and 1.9 that the Schur subgroup $S(k)$ of $Br(k)$ is, actually, a group.

We have shown that the group algebra $k[G]$ has an involution $*$ defined by (1.3). If the field k has an automorphism ι of order 2 ($\iota^2 = 1$), then $k[G]$ has another involution ι^* defined by

$$\iota^*: \sum_{g\in G} a_g g \to \sum_{g\in G} a_g^{\iota} g^{-1}, \quad (a_g \in k). \tag{1.5}$$

The most important case of this involution is that k is an imaginary abelian finite extension of the rationals Q. Then the automorphism ι of C over R induces an automorphism of k, denoted also by ι, whose fixed field is the maximal real subfield k_0 of k with $[k : k_0] = 2$. Let χ be an irreducible character of G with $k(\chi) = k$. Then, $\iota(\chi(g^{-1})) = \chi(g)$, and so

$$\begin{aligned}\iota^*(e(\chi)) &= \iota^*(\chi(1)|G|^{-1}(\sum_{g\in G} \chi(g^{-1})g)) \\ &= \chi(1)|G|^{-1}(\sum_{g\in G} \chi(g)g^{-1}) = e(\chi).\end{aligned}$$

This implies that the involution ι^* of $k[G]$ induces an involution of $k[G]e(\chi)$ (over $k_0 \cdot e(\chi) \cong k_0$) of the second-kind, in the sense of Albert [1]. Thus we have

Proposition 1.10. Let k be an abelian imaginary finite extension of Q and k_0 the maximal real subfield of k. Let $\langle\iota\rangle = G(k/k_0)$. Let χ be an irreducible character of G with $k(\chi) = k$. Then $A(\chi, k) = k[G]e(\chi)$ has an involution ι^* of the second kind (over k_0), defined by

$$\iota^*: \ (\sum_g a_g g)e(\chi) \to (\sum_g a_g^{\iota} g^{-1})e(\chi), \quad (a_g \in k). \tag{1.6}$$

Finally we observe the following.

Proposition 1.11. For every positive integer n, the complete matrix ring $M_n(k)$ is a Schur algebra over k.

Proof. From the formula for degree of an irreducible representation of a symmetric group (for instance, see [7]), it follows that the symmetric group S_n on n symbols ($n > 1$) has an irreducible character χ_n of degree $n - 1$. Since χ_n is realized in the rational field Q, it follows that $A(\chi_n, Q) \simeq M_{n-1}(Q)$, and $A(\chi_n, k) \simeq M_{n-1}(k)$. #

Corollary 1.12. Let D be a division algebra central over k such that $[D] \in S(k)$. Let t be the smallest positive integer such that $M_t(D)$ is a Schur algebra. Then, $M_{tn}(D)$ ($n = 1, 2, \cdots$) is a Schur algebra over k.

Proof. This is clear, because $M_{tn}(D) \simeq M_t(D) \otimes_k M_n(k)$. #

Let $A = A(\chi, k)$ be a Schur algebra over k, where χ is an irreducible character of a group G such that $k(\chi) = k$. Then $A \simeq A(\chi, Q(\chi)) \otimes_{Q(\chi)} k$. Hence we will mainly study the Schur subgroup for a cyclotomic field.

Chapter 2. CYCLOTOMIC ALGEBRAS

Throughout this chapter k denotes a field of characteristic 0. A <u>cyclotomic algebra</u> (Kreisalgebra) over k is a crossed product

$$B = (\beta,\ k(\zeta)/k) = \sum_{\sigma \varepsilon G} k(\zeta)u_\sigma \quad \text{(direct sum)}, \tag{2.1}$$

$$u_\sigma x = x^\sigma u_\sigma \ (x \ \varepsilon\ k(\zeta)), \quad u_\sigma u_\tau = \beta(\sigma,\ \tau)u_{\sigma\tau} \ (\sigma,\ \tau \ \varepsilon\ G),$$

where ζ is a root of unity, G is the Galois group of $k(\zeta)$ over k, and β is a factor set whose values are roots of unity in $k(\zeta)$. Note that G is abelian. It is clear that ζ and the values of β generate a finite cyclic group $<\zeta'>$ in $k(\zeta)^*$ and $k(\zeta') = k(\zeta)$, where ζ' is some root of unity. Hence we may assume $\zeta = \zeta'$. The Galois group G can be regarded as an automorphism group of the cyclic group $<\zeta>$ and the values of the factor set β belong to $<\zeta>$. By the theory of group extension (see, for instance, [57,III, §6]), it is easy to see that ζ and the elements u_σ $(\sigma \ \varepsilon\ G)$ generate a finite subgroup G in the multiplicative group B^*. Namely, G has a normal cyclic subgroup $<\zeta>$, the factor group $\mathrm{G}/<\zeta>$ is isomorphic to G, and β is exactly a factor set of the extension G of $<\zeta>$ by G. Since G spans B with coefficients in k, the center of B, it follows that B is a Schur algebra over k. Thus we have

<u>Proposition 2.1</u>. A cyclotomic algebra over k is a Schur algebra over k.

The u_σ $(\sigma \in G)$ are representatives of G over $\langle\zeta\rangle$. We may always assume that the factor set β is normalized, that is, $\beta(\sigma, 1) = \beta(1, \sigma) = u_1 = 1$ for any σ of G. Let the Galois group G be the direct product

$$G = \langle\sigma_1\rangle \times \langle\sigma_2\rangle \times \cdots \times \langle\sigma_r\rangle \tag{2.2}$$

of cyclic groups $\langle\sigma_i\rangle$ of order n_i. For an element σ of G, write

$$\sigma = \sigma_1^{\nu_1(\sigma)}\sigma_2^{\nu_2(\sigma)}\cdots\sigma_r^{\nu_r(\sigma)}, \quad 0 \leqq \nu_i(\sigma) < n_i \quad (i = 1,2,\cdots,r), \tag{2.3}$$

and put

$$v_\sigma = u_{\sigma_1}^{\nu_1(\sigma)}u_{\sigma_2}^{\nu_2(\sigma)}\cdots u_{\sigma_r}^{\nu_r(\sigma)}, \tag{2.4}$$

where $\nu_1(\sigma), \nu_2(\sigma), \cdots, \nu_r(\sigma)$ are integers uniquely determined by σ. Since $v_\sigma\zeta v_\sigma^{-1} = \zeta^\sigma = u_\sigma\zeta u_\sigma^{-1}$, it is easy to see that $v_\sigma u_\sigma^{-1}$ commutes with ζ. As $\langle\zeta\rangle$ is its own centralizer in G, there exists an element ω_σ in $\langle\zeta\rangle$ such that $v_\sigma = \omega_\sigma u_\sigma$. This implies that the elements v_σ $(\sigma \in G)$ defined by (2.3), (2.4) are also representatives of the group G over $\langle\zeta\rangle$ and give rise to a factor set β' equivalent (cohomolohous) to β:

$$\beta'(\sigma, \tau) = \omega_\sigma \omega_\tau^\sigma \omega_{\sigma\tau}^{-1} \beta(\sigma, \tau), \quad (v_\sigma = \omega_\sigma u_\sigma), \tag{2.5}$$

$$v_\sigma v_\tau = \beta'(\sigma, \tau) v_{\sigma\tau}. \tag{2.6}$$

We thus have

$$B = (\beta, k(\zeta)/k) = \sum_{\sigma \varepsilon G} k(\zeta) u_\sigma$$

$$= (\beta', k(\zeta)/k) = \sum_{\sigma \varepsilon G} k(\zeta) v_\sigma = \sum_{\nu_1=0}^{n_1-1} \cdots \sum_{\nu_r=0}^{n_r-1} k(\zeta) u_{\sigma_1}^{\nu_1} \cdots u_{\sigma_r}^{\nu_r}. \tag{2.7}$$

Note that

$$v_{\sigma\tau} = u_{\sigma_1}^{\nu_1(\sigma\tau)} u_{\sigma_2}^{\nu_2(\sigma\tau)} \cdots u_{\sigma_r}^{\nu_r(\sigma\tau)}, \tag{2.8}$$

$$0 \leqq \nu_i(\sigma\tau) < n_i, \quad \nu_i(\sigma\tau) \equiv \nu_i(\sigma) + \nu_i(\tau) \pmod{n_i},$$

$(i = 1, 2, \cdots, r)$. Put

$$h_i = u_{\sigma_i}^{n_i} = \beta(\sigma_i, \sigma_i)\beta(\sigma_i^2, \sigma_i) \cdots \beta(\sigma_i^{n_i-1}, \sigma_i), \quad (u_1 = 1), \tag{2.9}$$

$$h_{i,j} = u_{\sigma_i} u_{\sigma_j} u_{\sigma_i}^{-1} u_{\sigma_j}^{-1} = \beta(\sigma_i, \sigma_j)/\beta(\sigma_j, \sigma_i), \tag{2.10}$$

where $i, j = 1, 2, \cdots, r$; $i \neq j$. Then it is easy to see that for any σ, τ of G, $\beta'(\sigma, \tau)$ is of the form:

$$\beta'(\sigma, \tau) = (\prod_i h_i^{\xi_i})(\prod_{i<j} h_{j,i}^{\xi_{j,i}}), \tag{2.11}$$

where ξ_i and $\xi_{j,i}$ are some elements of the integral group ring $Z[G]$ (depending on σ and τ). This implies that the structure of both the group G and the cyclotomic algebra B is determined by the elements h_i and $h_{i,j}$. We observe

$$\beta'(\sigma_j, \sigma_i) = u_{\sigma_j} u_{\sigma_i} u_{\sigma_j}^{-1} u_{\sigma_i}^{-1}, \quad \beta'(\sigma_i, \sigma_j) = 1, \quad (1 \leqq i < j \leqq r), \tag{2.12}$$

$$\beta'(\sigma_i^{n_i - 1}, \sigma_i) = u_{\sigma_i}^{n_i}, \quad \beta'(\sigma_i^{\mu}, \sigma_i) = 1 \quad (0 \leqq \mu < n_i - 1). \tag{2.13}$$

It follows that for $i, j = 1, 2, \cdots, r$, $(i \neq j)$,

$$\begin{aligned} \beta'(\sigma_i, \sigma_j)/\beta'(\sigma_j, \sigma_i) &= u_{\sigma_i} u_{\sigma_j} u_{\sigma_i}^{-1} u_{\sigma_j}^{-1} = \\ h_{i,j} &= \beta(\sigma_i, \sigma_j)/\beta(\sigma_j, \sigma_i), \end{aligned} \tag{2.14}$$

$$\begin{aligned} \beta'(\sigma_i, \sigma_i)\beta'(\sigma_i^2, \sigma_i) \cdots \beta'(\sigma_i^{n_i - 1}, \sigma_i) &= u_{\sigma_i}^{n_i} = \\ h_i &= \beta(\sigma_i, \sigma_i)\beta(\sigma_i^2, \sigma_i) \cdots \beta(\sigma_i^{n_i - 1}, \sigma_i). \end{aligned} \tag{2.15}$$

Because ζ and $u_{\sigma_1}, u_{\sigma_2}, \cdots, u_{\sigma_r}$ generate a finite group G, which is an extension of the normal cyclic subgroup $<\zeta>$ by an abelian group isomorphic to $\mathcal{G} = \mathcal{G}(k(\zeta)/k)$, it follows that the elements h_i, $h_{i,j}$ $(i,j = 1, 2, \cdots, r;\ i \neq j)$ must satisfy the following relations:

$$h_i^{\sigma_i} = h_i, \tag{2.16}$$

$$h_{i,j} = h_{j,i}^{-1}, \quad (i < j), \tag{2.17}$$

$$h_j^{\sigma_i - 1} = h_{i,j}^{1 + \sigma_j + \cdots + \sigma_j^{n_j - 1}}, \quad (i \neq j), \tag{2.18}$$

$$h_{i,j}^{\sigma_\ell - 1} h_{j,\ell}^{\sigma_i - 1} h_{\ell,i}^{\sigma_j - 1} = 1, \quad (i < j < \ell). \tag{2.19}$$

Conversely, let ζ be a root of unity and let $G = G(k(\zeta)/k)$ be as in (2.2). Let h_i, $h_{i,j}$ $(i,j = 1, 2, \cdots, r;\ i \neq j)$ be roots of unity in $k(\zeta)$, which satisfy the above equations (2.16)-(2.19). We may assume that $h_i, h_{i,j} \in \langle\zeta\rangle$. Then G is regarded as an automorphism group of the cyclic group $\langle\zeta\rangle$, and the group G generated by $\zeta, u_{\sigma_1}, \cdots, u_{\sigma_r}$ with the defining relations

$$\text{(a)} \quad u_{\sigma_i} \zeta u_{\sigma_i}^{-1} = \zeta^{\sigma_i}, \tag{2.20}$$

$$\text{(b)} \quad u_{\sigma_i}^{n_i} = h_i, \tag{2.21}$$

$$\text{(c)} \quad u_{\sigma_i} u_{\sigma_j} u_{\sigma_i}^{-1} u_{\sigma_j}^{-1} = h_{i,j}, \quad (i < j) \tag{2.22}$$

contains the normal cyclic subgroup $\langle\zeta\rangle$ such that the factor group is the direct product of the cyclic groups $\langle\langle\zeta\rangle u_{\sigma_i}\rangle$ $(i = 1, 2, \cdots, r)$ of order n_i. That is, G is an extension of $\langle\zeta\rangle$ by G. (See [57, III, §8, Theorem 22].) If we write an element σ of G as in (2.3), then the elements v_σ $(\sigma \in G)$, given by (2.4), are representatives of the extension G over its normal subgroup $\langle\zeta\rangle$, and give rise to the factor set β' defined by (2.6) and of the form (2.11). In particular, the values of β' are in $\langle\zeta\rangle$. The factor set β' is also regarded as a factor set of the Galois group $G = G(k(\zeta)/k)$

with values in $k(\zeta)^*$ and defines a cyclotomic algebra over k:

$$(\beta', k(\zeta)/k) = \sum_{\sigma\varepsilon G} k(\zeta)v_\sigma = \sum_{\nu_1=0}^{n_1-1} \cdots \sum_{\nu_r=0}^{n_r-1} k(\zeta)u_{\sigma_1}^{\nu_1}\cdots u_{\sigma_r}^{\nu_r}.$$

Thus, in order to construct a cyclotomic algebra over k, we need only find some roots of unity h_i, $h_{i,j}$ satisfying (2.16)-(2.19).

Chapter 3. THE BRAUER-WITT THEOREM

In this chapter k is a field of characteristic 0. A group H is <u>k-elementary with respect to the prime</u> p if

(i) $H = AP$ (semi-direct product), where A is a cyclic, normal subgroup of H whose order is relatively prime to p, and P is a p-group.

(ii) Let $A = \langle a \rangle$, and let ζ be a primitive $|A|$-th root of unity. If a^i is conjugated to a^j in H, then there exists $\sigma \in G(k(\zeta)/k)$ such that $\sigma(\zeta^i) = \zeta^j$.

A group H is said to be k-<u>elementary</u> if it is k-elementary with respect to some prime p.

Let G be a group. A k-<u>representation</u> U <u>of</u> G is a homomorphism of G into the group of non-singular linear transformations of some finite-dimensional vector space V over k. A character of G which is afforded by a k-representation is said to <u>be realized</u> in k.

$Ch(G, k)$ is the ring of all Z-linear combinations of characters afforded by k-representations of G.

$V(G, k)$ (respectively, $V(G, k, p)$) is the ring of all Z-linear combinations of functions θ^G, where $\theta \in Ch(H, k)$ for some k-elementary subgroup H (respectively, for some k-elementary subgroup H with respect to the prime p) of G.

<u>Theorem 3.1</u>. (The fundamental theorem of character theory) We have

$$Ch(G, k) = V(G, k).$$

The heart of proof of the theorem is in the following

Proposition 3.2. Let $|G| = p^c b$, where $(p, b) = 1$. Then $b1_G \in V(G, k, p)$.

Proof. See [21, §15] or [15, §42].

Proof of Theorem 3.1. Let $\{p_i\}$ be the set of distinct primes dividing $|G|$. For each i let $|G| = p_i^{c_i} b_i$, where $(p_i, b_i) = 1$. By Proposition 3.2 $b_i 1_G \in V(G, k)$. There exist rational integers a_i such that $\Sigma\, a_i b_i = 1$. Thus

$$1_G = \sum_i a_i b_i 1_G \in V(G, k).$$

Since $V(G, k)$ is an ideal of $Ch(G, k)$, the theorem follows.

If χ is an irreducible character of G, and $G(k(\chi)/k) = \{\sigma_1, \sigma_2, \cdots, \sigma_r\}$, then we let $Tr_{k(\chi)/k}(\chi) = \chi^{\sigma_1} + \chi^{\sigma_2} + \cdots + \chi^{\sigma_r}$. Let U be a k-representation of G and ϕ its character. Then there are irreducible characters χ_i of G, which are not algebraically conjugate to each other over k, such that

$$\phi = \sum_i a_i m_k(\chi_i) Tr_{k(\chi_i)/k}(\chi_i),$$

a_i being positive integers (uniquely determined by ϕ).

Let L be an extension of k. Let ψ be the character of an L-representation of G such that $k(\psi) = k$. Then there exist irreducible characters χ_i of G, which are not algebraically conjugate to each other over k, such that

$$\psi = \sum_i b_i Tr_{k(\chi_i)/k}(\chi_i),$$

b_i being some positive integers.

Proposition 3.3. Let G be a group of exponent n and χ an irreducible character of G with $k(\chi) = k$. Let L be a field such that $k \subset L \subset k(\zeta)$, where ζ is a primitive n-th root of unity and $[k(\zeta) : L]$ is a power of p. Then there exists a subgroup H of G which is L-elementary with respect to p and an irreducible character ξ of H such that $L(\xi) = L$ and $(\chi, \xi^G) = (\chi|H, \xi) \not\equiv 0$ (mod p).

Proof. Let $|G| = p^c b$, where $(p, b) = 1$. By Proposition 3.2 there exist subgroups H_j of G which are L-elementary with respect to p such that $b1_G = \Sigma\, a_j \phi_j^G$, $(a_j \in Z)$, where ϕ_j is a character of H_j and $L(\phi_j) = L$. It follows that $L((\chi|H_j)\cdot\phi_j) = L$ and

$$b\chi = \Sigma\, a_j \chi\phi_j^G = \Sigma\, a_j((\chi|H_j)\cdot\phi_j)^G.$$

For any irreducible character ξ of H_j let $T(\xi) = Tr_{L(\xi)/L}(\xi)$. Thus there exist subgroups H_i of G which are L-elementary with respect to p and irreducible characters ξ_i of H_i such that $b\chi = \Sigma\, d_i T(\xi_i)^G$. Hence

$$b = (\chi, b\chi) = \Sigma\, d_i(\chi, T(\xi_i)^G)$$

$$= \Sigma\, d_i(\chi|H_i, T(\xi_i)) = \Sigma\, d_i[L(\xi_i) : L](\chi|H_i, \xi_i).$$

Since $(b, p) = 1$ this implies that for some $\xi = \xi_i$ and $H = H_i$

$$[L(\xi) : L](\chi|H, \xi) \not\equiv 0 \pmod p.$$

Since $L(\xi) \subset k(\zeta)$ and $[k(\zeta) : L]$ is a power of p this implies that $L(\xi) = L$ and $(\chi|H, \xi) \not\equiv 0$ (mod p) as required. #

Let $N \lhd G$ and let ψ be a character of N. For $g \in G$, ψ^g

is the character of N defined by $\psi^g(n) = \psi(gng^{-1})$, $n \in N$.

Proposition 3.4. Let $N \lhd G$. Let χ be an irreducible character of G which is induced by an irreducible character ψ of N, and such that $k(\chi) = k$. Set $F = \{f \in G;\ \psi^f = \psi^{\tau(f)}$ for some $\tau(f) \in G(k(\psi)/k)\}$. Let Nf_i $(i = 1,2,\cdots,t;\ f_1 = 1)$ be all the distinct cosets of N in F, and set $\tau(f_i) = \tau_i$, $(\tau_1 = 1)$. Then, (i) $F/N \simeq \{\tau_1, \tau_2, \cdots, \tau_t\} = G(k(\psi)/k)$, (ii) $k(\psi^F) = k$.

Proof. It is clear that the mapping: $f \to \tau(f)$ is a homomorhhism of F into $G(k(\psi)/k)$, whose kernel contains N. Set $\theta = \psi^F$. Since $N \lhd F$, it follows that $\theta(f) = 0$ for $f \notin N$ and that

$$\theta(n) = \sum_{i=1}^{t} \psi^{f_i}(n) = \sum_{i=1}^{t} \psi^{\tau_i}(n) \quad \text{for } n \in N.$$

Since θ is an irreducible character of F, we have

$$1 = (\theta, \theta)_F = \frac{1}{|F|}\sum_{f \in F} \theta(f)\theta(f^{-1})$$

$$= \frac{|N|}{|F|}\frac{1}{|N|}\sum_{n \in N}\Big(\sum_{i=1}^{t}\psi^{\tau_i}(n)\Big)\Big(\sum_{j=1}^{t}\psi^{\tau_j}(n^{-1})\Big) = \frac{1}{t}\sum_{i=1}^{t}\sum_{j=1}^{t}(\psi^{\tau_i}, \psi^{\tau_j})_N.$$

This yields that $(\psi^{\tau_i}, \psi^{\tau_j}) = 0$ for $i \neq j$, and that the τ_i are distinct. Thus, $F/N \simeq \{\tau_1, \tau_2, \cdots, \tau_t\} \subset G(k(\psi)/k)$. Since $(\theta(n))^{\tau_j} = \sum_i \psi^{\tau_i\tau_j}(n) = \theta(n)$, $n \in N$, and $\theta(f) = 0$, $f \notin N$, it follows that $\tau_j \in G(k(\psi)/k(\theta))$, $(j = 1,2,\cdots,t)$. Conversely, for any $\tau \in G(k(\psi)/k(\theta))$,

$$\sum_{i=1}^{t}\psi^{\tau_i}(n) = \theta(n) = \theta^{\tau}(n) = \sum_{i=1}^{t}\psi^{\tau_i\tau}(n), \quad n \in N.$$

Hence $\tau = \tau_i$ for some i. Thus, $\{\tau_1, \tau_2, \cdots, \tau_t\} = G(k(\psi)/k(\theta))$.

Note that $k(\psi) \supset k(\theta) = k(\psi^F) \supset k = k(\chi) = k(\psi^G)$. If $\tau \in G(k(\psi)/k)$, then $(\psi^G)^\tau = \psi^G$, and so

$$(\psi^G)^\tau(n) = \sum_{i=1}^{s} (\psi^{g_i})^\tau(n) = \sum_{i=1}^{s} \psi^{g_i}(n) = \psi^G(n), \quad n \in N,$$

where $Ng_1, Ng_2, \cdots, Ng_s$ $(g_1 = 1)$ are all the distinct cosets of N in G. This implies that $(\psi^{g_1})^\tau = \psi^\tau = \psi^{g_i}$ for some i. It then follows from the definition of the subgroup F that $g_i \in F$ and $\tau \in \{\tau_1, \tau_2, \cdots, \tau_t\}$. Thus we conclude that $k(\theta) = k$, establishing the proposition.

<u>Proposition 3.5</u>. Let the notation and assumption be as in Proposition 3.4. Let $f_i f_j = n_{ij} f_{\nu(i,j)}$, $n_{ij} \in N$, $\nu(i,j) \in \{1, 2, \cdots, t\}$. Suppose that ψ is a linear character of N, and put $\beta(\tau_i, \tau_j) = \psi(n_{ij})$, $(1 \leqq i, j \leqq t)$. Then, β is a factor set of $G(k(\psi)/k)$ consisting of roots of unity, and the Schur algebra $A(\psi^F, k)$ over k is k-isomorphic to the cyclotomic algebra $(\beta(\tau_i, \tau_j), k(\psi)/k)$ over k.

<u>Proof</u>. Define $U(f) = (\psi(f_i f f_j^{-1}))$ for $f \in F$, where $U(f)$ is a matrix of degree t, having $\psi(f_i f f_j^{-1})$ in the i-th row and j-th column. Then U is an irreducible representation of F, induced by ψ and with the character ψ^F. For $n \in N$, $U(n)$ is the diagonal matrix $[\psi^{\tau_1}(n), \cdots, \psi^{\tau_t}(n)]$, having $\psi^{\tau_i}(n)$ in the i-th row and i-th column $(\tau_1 = 1)$. Put $\Xi = \mathrm{env}_k(U|N)$, where $U|N$ is the restriction of U to N. It follows that

$$\Xi = \{[x^{\tau_1}, \cdots, x^{\tau_t}];\ x \in k(\psi)\}. \tag{3.1}$$

Hence by the mapping $x \to [x^{\tau_1}, \cdots, x^{\tau_t}]$, $(x \in k(\psi))$, $k(\psi)$ is isomorphic to Ξ. It is easy to see that

$$\mathrm{env}_k(U) = \sum_{i=1}^{t} \Xi \cdot U(f_i), \tag{3.2}$$

and that the mapping

$$\tau_i' : \ X \to U(f_i)XU(f_i)^{-1}, \quad X \in \mathrm{env}_k(U|N) \tag{3.3}$$

is the automorphism of the field Ξ ($\simeq k(\psi)$) over $k \cdot 1_t$ ($\simeq k$), which corresponds to $\tau_i \in G(k(\psi)/k)$, $(i = 1,2,\cdots,t)$. So, $\{\tau_1', \tau_2', \cdots, \tau_t'\} = G(\Xi/k\cdot 1_t)$, and the matrices $U(f_1), U(f_2), \cdots, U(f_t)$ are linearly independent over the field Ξ (cf. [40, §139]). Thus the central simple algebra $\mathrm{env}_k(U)$ over $k\cdot 1_t$ is expressed as a crossed product:

$$\mathrm{env}_k(U) = \sum_{i=1}^{t} \Xi\cdot U(f_i) = (\beta'(\tau_i', \tau_j'), \Xi/k\cdot 1_t),$$

$$U(f_i)U(f_j) = \beta'(\tau_i', \tau_j')U(f_{\nu(i,j)}), \quad \beta'(\tau_i', \tau_j') = U(n_{ij}),$$

$$U(f_i)XU(f_i)^{-1} = X^{\tau_i'}, \quad (X \in \Xi), \quad (1 \leq i, j \leq n).$$

By associativity, β' is a factor set of $G(\Xi/k\cdot 1_t)$. By the isomorphism: $\Xi \simeq k(\psi)$, $\beta'(\tau_i', \tau_j')$ corresponds to $\psi(n_{ij}) = \beta(\tau_i, \tau_j)$. We thus conclude that $\mathrm{env}_k(U) \simeq (\beta(\tau_i, \tau_j), k(\psi)/k)$. If N' denotes the kernel of ψ and ζ is a primitive $|N/N'|$-th root of unity, then $k(\psi) = k(\zeta)$ and $\beta(\tau_i, \tau_j) \in \langle\zeta\rangle$. Hence the above crossed product is a cyclotomic algebra over k. Finally we recall that $\mathrm{env}_k(U) \simeq A(\psi^F, k)$, (cf. Proposition 1.3). #

It is clear that a k-elementary group is also a Q-elementary group, Q being the rationals. Let $H = \langle a\rangle P$ be a Q-elementary group with respect to p, where $H \rhd \langle a\rangle$, P is a p-group, and $(|\langle a\rangle|, p) = 1$. Let ξ be an irreducible character of H. It is known that ξ is monomial and the degree of ξ is a power of p. (See, for instance, [21, 10.2 and 9.13] or [30, V, 18.4 and 17.10].) We see that every subgroup F of H whose index $[H : F]$ is a power of p necessarily contains the normal subgroup $\langle a\rangle$. Hence, if H is k-elementary, then F is also k-elementary. Furthermore, if ξ is induced by a linear character of a subgroup H', then H' contains $\langle a\rangle$.

<u>Proposition 3.6</u>. Let $H = \langle a\rangle \cdot P$ be a k-elementary group with respect to the prime p, and ξ an irreducible character of H such that $k(\xi) = k$. Then there exist subgroups F, N of H and a linear character ψ of N such that (i) $F \rhd N \supset \langle a\rangle$, and F is k-elementary with respect to p, (ii) $\xi = \psi^H$, (iii) for each $f \in F$, $\psi^f = \psi^{\tau(f)}$ for some $\tau(f) \in G(k(\psi)/k)$, (iv) $k(\psi^F) = k$.

<u>Proof</u>. Let T be a minimal normal subgroup of H such that i) there exists a character θ of T from which ξ is induced, ii) for each $h \in H$, $\theta^h = \theta^{\tau(h)}$ for some $\tau(h) \in G(k(\theta)/k)$. Assume first that θ is a linear character of T. Then by setting $F = H$, $N = T$ and $\psi = \theta$, all the statements of the proposition hold. Assume next that θ is non-linear. We note that a subgroup of index p of a k-elementary group with respect to p is normal. From the remarks preceding Proposition 3.6 it follows that T is

a k-elementary group with respect to p containing $\langle a\rangle$, and that there exists a subgroup S of T and a character ρ of S such that $[T : S] = p$ and $\rho^T = \theta$. As S is normal in T and ρ induces θ, we see that θ vanishes outside S. Let U be the intersection of all conjugates hSh^{-1} $(h \in H)$ of S in H. Since $\langle a\rangle \subset S$ and $\langle a\rangle \lhd H$, it follows that $\langle a\rangle \subset U$. So $U \lhd H$ and H/U is a p-group. Hence there exists $Y \lhd H$ such that $U \subset Y \subset T \subset H$ and $[T : Y] = p$. Since for every $h \in H$, θ^h is an algebraic conjugate of θ and θ vanishes outside S, it follows that θ vanishes outside U, a fortiori, outside Y. From the equation

$$(\theta|Y, \theta|Y)_Y = |Y|^{-1} \sum_{y \in Y} \theta(y)\theta(y^{-1}) = p|T|^{-1} \sum_{t \in T} \theta(t)\theta(t^{-1}) = p,$$

we conclude that there exists an irreducible character ϕ of Y such that $\theta|Y = \sum_x \phi^x$, where x ranges over a complete system of coset representatives of Y in T and the characters ϕ^x $(x \in T \bmod Y)$ are distinct from each other. Namely the inertial group of ϕ is Y and the index of ramification of θ with respect to Y is equal to 1 (cf. [21, §9]). Thus ϕ induces θ and so ξ is induced by ϕ. Let E be the subgroup of H consisting of all $h \in H$ such that $\phi^h = \phi^{\tau(h)}$ for some $\tau(h) \in G(k(\phi)/k)$. By minimality of T, E is a proper subgroup of H containing $\langle a\rangle$. By Proposition 3.4, we conclude that $k(\phi^E) = k$. Thus we have proved that if θ is non-linear, then there exists a proper subgroup E of H which contains $\langle a\rangle$ and hence is k-elementary, and a character η (= ϕ^E) of E such that $\eta^H = \xi$ and $k(\eta) = k$. Using these arguments successively, we are able to find subgroups F, N, and a linear character ψ of N satisfying the conditions (i)-(iv) of Proposition 3.6. #

Here we summarize well-known facts about the Brauer group $Br(k)$. If A is a central simple algebra over k, there exists a finite Galois extension L of k which splits A. Then A is similar to a crossed product of L/k with a factor set α:

$$\alpha(\sigma, \tau)^{\rho}\alpha(\rho, \sigma\tau) = \alpha(\rho, \sigma)\alpha(\rho\sigma, \tau), \quad \rho, \sigma, \tau \in G(L/k),$$

which is nothing but a 2-cocycle of G with values in L^*. The element of the 2-cohomology group $H^2(L/k)$, which is determined by α, is also denoted by α. By this correspondence the subgroup $Br(L/k)$ of $Br(k)$ consisting of those algebra classes $[A]$ which are split by L is identified with $H^2(L/k)$, and the Brauer group $Br(k)$ is identified with the inductive limit of $H^2(L/k)$, where L ranges over all the finite Galois extensions of k. Let E be an extension of k with $[E : k] = n$. Denote by Res the restriction homomorphism of $Br(k)$ into $Br(E)$. It follows from [16, V, §4, Satz 1] that for $[A] \in Br(k)$, $Res([A]) = [A \otimes_k E]$. In particular, if χ is an irreducible character of a group G with $k(\chi) = k$, then $Res([A(\chi, k)]) = [A(\chi, E)]$. Denote by Cor the corestriction homomorphism of $Br(E)$ into $Br(k)$. Then it is well-known that $Cor \circ Res([A]) = ([A])^n$ (cf. [35, VII, §7, Proposition 6]). Let p be a prime such that $(p, n) = 1$. Let

$Br(k)_p$ denote the subgroup of $Br(k)$ consisting of all the classes $[A]$ of $Br(k)$ whose exponents are powers of p. Then the restriction homomorphism, $Res: Br(k) \to Br(E)$, induces a homomorphism of $Br(k)_p$ into $Br(E)_p$, denoted also by Res.

Lemma 3.7. Let p be a prime. Let E be an extension of k of degree n with $(n, p) = 1$. Then the homomorphism Res: $Br(k)_p \to Br(E)_p$ is injective.

Proof. Let $[A] \in Br(k)_p$. Suppose that $Res([A]) = [A \otimes_k E] = [E]$. Then the index of A divides $n = [E : k]$, and $(n, p) = 1$. Since the exponent of A is a power of p and divides the index of A, it follows that the index of A is equal to 1 and $[A] = [k]$. The assertion now follows immediately. #

Let D be a division algebra central over k with the index m. If $m = p_1^{a_1} p_2^{a_2} \cdots p_s^{a_s}$ is the factorization of m into prime powers, then D is k-isomorphic to the product $D_1 \otimes D_2 \otimes \cdots \otimes D_s$, where D_i is a division algebra central over k with the index $p_i^{a_i}$ ([16, V, §3, Satz 3]). Furthermore, D_i is uniquely determined by D. We call the algebra class $[D_i]$ the p_i-part of $[D]$, and denote it by $[D]_{p_i}$. If $p \nmid m$, let the p-part $[D]_p$ of $[D]$ be equal to $[k]$.

Proposition 3.8. Let χ be an irreducible character of G with $k(\chi) = k$. Let H be a subgroup of G and ξ an irreducible character of H such that $k(\xi) = k$, and that $(\chi|H, \xi) = t \neq 0$. Then, for each prime p with $(p, t) = 1$, the p-parts of $[A(\chi, k)]$ and $[A(\xi, k)]$ are the same.

Proof. Let n denote the exponent of G. The field $k(\zeta_n)$ is a splitting field for both $A(\chi, k)$ and $A(\xi, k)$. Let L be the subfield of $k(\zeta_n)$ over k such that $[k(\zeta_n) : L]$ is a power of p and $[L : k] \not\equiv 0 \pmod p$. Then the exponents of $[A(\chi, L)]$ and $[A(\xi, L)]$ in $Br(L)$ are both powers of p. It follows from Propositions 1.9 and 1.7 that $[A(\chi \otimes \bar{\chi}, L)] = [A(\chi, L)]\cdot[A(\bar{\chi}, L)] = [L]$, $(\bar{\chi}(g) = \chi(g^{-1}),\ g \in G)$. So $\chi \otimes \bar{\chi}$ is realized in L, whence the character $(\chi|H) \otimes \bar{\chi}$ of $H \times G$ is also realized in L. The character $\xi \otimes \bar{\chi}$ of $H \times G$ is irreducible and $L(\xi \otimes \bar{\chi}) = L$. It is easy to see that $((\chi|H) \otimes \bar{\chi}, \xi \otimes \bar{\chi})_{G\times H} = (\chi|H, \xi)_H \cdot (\bar{\chi}, \bar{\chi})_G = t$. Hence the Schur index $m_L(\xi \otimes \bar{\chi})$ divides t. Since $[A(\xi \otimes \bar{\chi}, L)] = [A(\xi, L)]\cdot[A(\bar{\chi}, L)]$, it follows that the exponent of $[A(\xi \otimes \bar{\chi}, L)]$, which divides $m_L(\xi\otimes \bar{\chi})$, is a power of p. Because $(p, t) = 1$, we conclude that $m_L(\xi \otimes \bar{\chi}) = 1$, whence $[A(\xi, L)] = [A(\bar{\chi}, L)]^{-1} = [A(\chi, L)]$. Then Lemma 3.7 yields that the p-parts of $[A(\xi, k)]$ and $[A(\chi, k)]$ are the same. #

Corollary 3.9. Let χ be an irreducible character of G with $k(\chi) = k$, and let ξ be a character of a subgroup H of G such that $k(\xi) = k$ and $\chi = \xi^G$. Then, $[A(\chi, k)] = [A(\xi, k)]$.

Proof. We have $(\chi|H, \xi) = (\chi, \xi^G) = (\chi, \chi) = 1$. Hence, for every prime p, $[A(\chi, k)]_p = [A(\xi, k)]_p$, whence $[A(\chi, k)] = [A(\xi, k)]$. #

Now we are ready to state

THE BRAUER-WITT THEOREM. Let k be a field of characteristic 0. Let G be a finite group of exponent n and χ an irreducible character of G with $k(\chi) = k$. Let p be a prime number.

(I) Let L be the subfield of $k(\zeta_n)$ over k such that $[k(\zeta_n) : L]$ is a power of p and $[L : k] \not\equiv 0 \pmod p$. Then there is a subgroup F of G which is L-elementary with respect to p, and an irreducible character θ of F with $L(\theta) = L$, such that $(\chi|F, \theta) \not\equiv 0 \pmod p$, and that the following statement (II) holds.

(II) There is a normal subgroup N of F and a linear character ψ of N such that (i) $\theta = \psi^F$, (ii) for each $f \in F$, there exists $\tau(f) \in G(L(\psi)/L)$ such that $\psi^f = \psi^{\tau(f)}$, and by the mapping $f \to \tau(f)$, $F/N \simeq G(L(\psi)/L)$, (iii) $A(\theta, L)$ is L-isomorphic to the cyclotomic algebra $(\beta, L(\psi)/L)$ over L, where, if T is a complete set of coset representatives of N in F ($1 \in T$) with $ff' = n(f, f')f''$ for $f, f', f'' \in T$, $n(f, f') \in N$, then $\beta(\tau(f), \tau(f')) = \psi(n(f, f'))$.

(III) $[A(\chi, L)] = [A(\theta, L)] = [(\beta, L(\psi)/L)]$ in $Br(L)$, and the p-part of $m_k(\chi)$ (i.e., the highest power of p dividing $m_k(\chi)$) is equal to $m_L(\theta)$.

(IV) Let K be a field of characteristic 0 such that $K(\chi) = K$. Let H be a subgroup of G, and ξ an irreducible character of H such that $K(\xi) = K$ and $(\chi|H, \xi) \not\equiv 0 \pmod p$. Then the p-parts of the classes $[A(\chi, K)]$ and $[A(\xi, K)]$ are the same.

Proof. By Proposition 3.3, there is a subgroup H of G which is L-elementary with respect to p and an irreducible character ξ of H with $L(\xi) = L$, such that $(\chi|H, \xi) = (\chi, \xi^G) \not\equiv 0 \pmod p$. Then Proposition 3.6 yields that there exists a subgroup F of H which is also L-elementary with respect to p, and that there is a subgroup N of F and a linear character ψ of N such that (1) $F \triangleright N$, (2) $\psi^H = (\psi^F)^H = \xi$, (3) for each $f \in F$, there exists $\tau(f) \in G(L(\psi)/L)$ such that $\psi^f = \psi^{\tau(f)}$, (4) $L(\psi^F) = L$. Putting $\theta = \psi^F$, we have $L(\theta) = L$, $\xi^G = (\theta^H)^G = \theta^G$, and $(\chi|F, \theta) = (\chi, \theta^G) = (\chi, \xi^G) \not\equiv 0 \pmod p$. This proves the statement (I). Furthermore, it follows easily from Proposition 3.4 that $F/N \cong G(L(\psi)/L)$, and it follows from Proposition 3.5 that $A(\theta, L) \approx (\beta, L(\psi)/L)$, where β is such a factor set of $L(\psi)/L$ as is described in the statement (II). Since $k(\zeta_n)$ is a splitting field for both $A(\chi, L)$ and $A(\theta, L)$, and $[k(\zeta_n) : L]$ is a power of p, we see that $[A(\chi, L)] = [A(\chi, L)]_p$ and $[A(\theta, L)] = [A(\theta, L)]_p$. Hence by Proposition 3.8, $[A(\chi, L)] = [A(\theta, L)] = [(\beta, L(\psi)/L)]$. The p-part of $m_k(\chi)$ is the index of $[A(\chi, k)]_p$, which is equal to the index of $[A(\chi, L)] = [A(\theta, L)]$, i.e., $m_L(\theta)$, establishing the statement (III). The assertion (IV) has been given in Proposition 3.8. The proof of the Brauer-Witt theorem is completed.

Corollary 3.10. Let the notation be as in the Brauer-Witt

theorem. Then $A(\chi, k)$ is similar to a cyclotomic algebra over k.

Proof. We have seen that $\mathrm{Res}([A(\chi, k)]_p) = [A(\chi, L)] = [B]$ where $B = (\beta, L(\psi)/L)$ is a cyclotomic algebra over L. It follows from the definition of corestriction homomorphism that $\mathrm{Cor}([B])$ is also a class of $\mathrm{Br}(k)$, which is represented by a cyclotomic algebra over k, where $\mathrm{Cor}\colon \mathrm{Br}(L) \to \mathrm{Br}(k)$. Put $[L : k] = t \not\equiv 0 \pmod p$ and $[k(\zeta_n) : L] = p^s$. Let r be an integer such that $rt \equiv 1 \pmod{p^s}$. Then $(\mathrm{Cor}([B]))^r = (\mathrm{Cor}\circ\mathrm{Res}([A(\chi, k)]_p))^r = ([A(\chi, k)]_p)^{tr} = [A(\chi, k)]_p$, because the exponent of $[A(\chi, k)]_p$ divides p^s. Since p is an arbitrary prime, we conclude that $[A(\chi, k)]$ is a class represented by a cyclotomic algebra over k. #

Corollary 3.11. The Schur subgroup $S(k)$ consists of all those algebra classes of $\mathrm{Br}(k)$ that contain a cyclotomic algebra over k.

Proof. This follows at once from Proposition 2.1 and Corollary 3.10.

Remark 3.12. In the statement (II) of the Brauer-Witt Theorem, we denote by N_0 the kernel of ψ. Then it is easy to see that N_0 is also the kernel of $\theta = \psi^F$, so $N_0 \triangleleft F$. Put $F/N_0 = F'$, $N/N_0 = N'$. We see that θ and ψ are regarded as faithful representations of F' and N' respectively, and $\theta = \psi^{F'}$, $(F/N \cong F'/N')$.

Chapter 4. THE SCHUR SUBGROUP OF A p-ADIC FIELD, $p \neq 2$

Let p be a prime, and Q_p the rational p-adic numbers. Let $K \supset k$ be finite extensions of Q_p. In Chapters 4 and 5, $f_{K/k}$ denotes the residue class degree of K/k, and $e_{K/k}$ denotes the ramification index of K/k. Let h be a positive integer such that $h = p^n h'$, $n \geq 0$, $(p, h') = 1$. Let f be the smallest positive integer such that $p^f \equiv 1 \pmod{h'}$. Then it is wellknown that $Q_p(\zeta_h) = Q_p(\zeta_{p^n}, \zeta_{p^f-1})$. Suppose that k is a cyclotomic extension of Q_p. Let $B = (\beta, k(\zeta)/k)$ be a cyclotomic algebra over k, where ζ is a root of unity and β is a factor set of $k(\zeta)/k$ whose values are roots of unity in $k(\zeta)$. Let $Q_p(\zeta')$ be a cyclotomic field containing $k(\zeta)$, where ζ' is some root of unity. Then

$$B = (\beta, k(\zeta)/k) \sim (\mathrm{Inf}(\beta), Q_p(\zeta')/k),$$

where Inf denotes the inflation map from $H^2(k(\zeta)/k)$ into $H^2(Q_p(\zeta')/k)$. Thus we may always assume that an arbitrary cyclotomic algebra over k is of the form:

$$B = (\beta, L/k) = \sum_{\sigma \varepsilon G} Lu_\sigma, \quad L = Q_p(\zeta), \tag{4.1}$$

$$u_\sigma u_\tau = \beta(\sigma, \tau)u_{\sigma\tau}, \quad u_\sigma x = x^\sigma u_\sigma \quad (x \varepsilon L), \quad (u_1 = 1) \tag{4.2}$$

where ζ is a root of unity, $G = G(L/k)$, and $\beta(\sigma, \tau)$ is a root of unity for every $\sigma, \tau \varepsilon G$. We will calculate the index of B. Let $\omega(L)$ denote the group of roots of unity contained in L. Let $\omega'(L)$ (respectively, $\omega_p(L)$) denote the subgroup of $\omega(L)$ consisting of those roots of unity in L whose orders are relatively prime to p (respectively, powers of p). We have

$$\omega(L) = \omega'(L) \times \omega_p(L), \tag{4.3}$$

$$\omega'(L) = \langle \zeta_{q^f-1} \rangle, \quad \omega_p(L) = \langle \zeta_{p^n} \rangle, \tag{4.4}$$

where $q = p^{f^*}$, $f^* = f_{k/Q_p}$ (the residue class degree of k/Q_p), $f = f_{L/k}$, and n is some integer $\geqq 0$. Then,

$$L = Q_p(\zeta) = Q_p(\zeta_{q^f-1}, \zeta_{p^n}). \tag{4.5}$$

According to (4.3),

$$\beta(\sigma, \tau) = \alpha(\sigma, \tau)\gamma(\sigma, \tau), \tag{4.6}$$

$$\alpha(\sigma, \tau) \in \omega'(L), \quad \gamma(\sigma, \tau) \in \omega_p(L), \tag{4.7}$$

for all $\sigma, \tau \in G$, and hence

$$B = (\beta, L/k) \sim (\alpha, L/k) \otimes_k (\gamma, L/k). \tag{4.8}$$

Now we will present a lemma which is one of the ideas for computing the index of a p-adic cyclotomic algebra.

Lemma 4.1 (Yamada [45]). Let $\Lambda \supset K$ be finite extensions of Q_p such that Λ/K is normal. Put $e = e_{\Lambda/K}$, $f = f_{\Lambda/K}$. Let z be a positive integer divisible by $ef = [\Lambda : K]$, and let Ω be the unramified extension of K of degree z. Set $\Lambda' = \Lambda\cdot\Omega$. Then $e_{\Lambda'/K} = e$, $f_{\Lambda'/K} = z$, and the order of every Frobenius automorphism of Λ'/K is equal to z. Furthermore, there is a totally ramified extension F of k in Λ' with $[F : k] = e$, so that $F\cdot\Omega = \Lambda'$ and $F \cap \Omega = K$. The inertia group of Λ'/K is canonically isomorphic to that of Λ/K.

Proof. Since an unramified extension is uniquely determined by its degree, it follows that $[\Omega \cap \Lambda : K] = f$. Hence $\Lambda' = \Lambda\cdot\Omega$ is normal over K of degree ze, Λ'/Ω is totally ramified of degree e, and Λ'/Λ is unramified of degree z/f. Set $G(\Lambda'/K) = G$, $G(\Lambda'/\Lambda) = H$, and $G(\Lambda'/\Omega) = H_1$. Then $H \cap H_1 = 1$, $|G/H| = ef$, and $|G/H_1| = z$. This implies that for any element σ of G, $\sigma^z H = (\sigma H)^z = H$, $\sigma^z H_1 = (\sigma H_1)^z = H_1$, and hence σ^z belongs to $H \cap H_1 = 1$, i.e., $\sigma^z = 1$. The assertions of the lemma easily follow.

Next we recall a wellknown fact.

Lemma 4.2. Let $E \supset K$ be finite extensions of Q_p. Suppose that E/K is unramified. Let $(\beta, E/K)$ be a crossed product of E/K with a factor set β, whose values are units in E. Then, $(\beta, E/K) \sim K$.

Proof. See, for instance, [51, Lemma 7].

For the rest of this chapter, we assume $p \neq 2$, and use the same notation as is given in (4.1)-(4.8). If $n = 0$ in (4.4), then L/k is unramified, and hence $B = (\beta, L/k) \sim k$ by Lemma 4.2. Therefore we always assume $n \geq 1$. The multiplicative group $Z \bmod^{\times} p^n$ is cyclic, and the inertia group T of L/k is isomorphic to a subgroup of $Z \bmod^{\times} p^n$. Put $e = |T| = e_{L/k}$. Let ω be a generator of T: $T = \langle\omega\rangle$, $\omega^e = 1$. Let η be a Frobenius automorphism of L/k. Then $G = G(L/k) = \langle\omega, \eta\rangle$. Denote by $\mathfrak{p}$ the prime ideal of k, so $N_{k/Q_p}(\mathfrak{p}) = q$.

<u>Theorem 4.3</u> (Yamada [45]). Let p be an odd prime and k a cyclotomic extension of Q_p. Let $B = (\beta, L/k)$ be a cyclotomic algebra over k given by (4.1)-(4.8). Notation being as before, let c be the index of tame ramification of k/Q_p, i.e., $p = \mathfrak{p}^{cp^d}$, $(c, p) = 1$ for some integer d. Then the number

$$\delta = (\alpha(\omega,\eta)/\alpha(\eta,\omega))^{e/(q-1)}\alpha(\omega,\omega)\alpha(\omega^2,\omega)\cdots\alpha(\omega^{e-1},\omega)$$

belongs to the field k, so that we can write $\delta = \zeta_{q-1}^{v}$ for a certain integer v. The index of the cyclotomic algebra $(\beta, L/k) \sim (\alpha, L/k) \otimes (\gamma, L/k)$ is equal to

$$\frac{(p - 1)/c}{(v, (p - 1)/c)}. \tag{4.9}$$

<u>Proof</u>. First we will calculate the index of the cyclotomic algebra $(\alpha, L/k)$, $(\alpha(\sigma, \tau) \in \langle\zeta_{q^f-1}\rangle$, $\sigma, \tau \in G(L/k))$. Let Ω be an unramified extension of k of degree z such that

(i) Ω contains a $(q-1)$-th root $(\alpha(\sigma, \tau))^{1/(q-1)}$ of $\alpha(\sigma, \tau)$

for every σ, τ of $G(L/k)$, and (ii) $ef \mid z$. Since an unramified extension is uniquely determined by its degree, it follows that $\Omega = k(\zeta_{q^z-1})$ and $\Omega\cdot L = Q_p(\zeta_{q^z-1}, \zeta_{p^n})$. Put $L' = \Omega\cdot L$. Then L' is an abelian extension of k of degree ze. Lemma 4.1 yields that there exists a totally ramified extension F of k in L' of degree $e = e_{L/k} = e_{L'/k}$, so that $F\cdot\Omega = L'$, $F \cap \Omega = k$. Furthermore, there are automorphisms θ, ϕ of L' over k such that (i) $\langle\theta\rangle = G(L'/\Omega)$, $\langle\phi\rangle = G(L'/F)$, $G(L'/k) = \langle\theta\rangle \times \langle\phi\rangle$, $\theta^e = \phi^z = 1$; (ii) $\langle\theta\rangle$ is the inertia group of L'/Ω and θ extends ω, i.e., $\theta|L = \omega$; (iii) ϕ is a Frobenius automorphism of L'/k, which extends η. Put $\alpha' = \mathrm{Inf}(\alpha)$, where Inf denotes the inflation homomorphism of $H^2(L/k)$ into $H^2(L'/k)$. Then

$$(\alpha', L'/k) \sim (\alpha, L/k).$$

So we will calculate the index of the cyclotomic algebra

$$B' = (\alpha', L'/k) = \sum_{\sigma\in G'} L'v_\sigma,$$

$$v_\sigma v_\tau = \alpha'(\sigma, \tau)v_{\sigma\tau}, \quad v_\sigma x = x^\sigma v_\sigma \quad (x \in L'),$$

where $G' = G(L/k)$ and $\sigma, \tau \in G'$. Since $G' = \langle\theta\rangle \times \langle\phi\rangle$, we have

$$B' = \sum_{i=0}^{e-1} \sum_{j=0}^{z-1} F\cdot\Omega v_\theta^i v_\phi^j.$$

Put

$$\lambda = (\alpha'(\theta, \phi)/\alpha'(\phi, \theta))^{1/(q-1)}.$$

Since $\theta|L = \omega$ and $\phi|L = \eta$, it follows from the definition of inflation map that

$$\lambda = (\alpha(\omega,\ \eta)/\alpha(\eta,\ \omega))^{1/(q-1)}.$$

By the assumption for the field Ω, λ belongs to $<\zeta_{q^z-1}> \subset \Omega$, and so $\lambda v_\theta = v_\theta\lambda$. Note that $\zeta^\phi_{q^z-1} = \zeta^q_{q^z-1}$. We have

$$v_\theta v_\phi = (\alpha'(\theta,\ \phi)/\alpha'(\phi,\ \theta))v_\phi v_\theta = \lambda^{q-1}v_\phi v_\theta,$$

$$(\lambda v_\theta)v_\phi = \lambda^q v_\phi v_\theta = v_\phi(\lambda v_\theta).$$

Since λv_θ commutes with v_ϕ, and each element of Ω (resp. F) commutes with λv_θ (resp. v_ϕ), it follows that

$$\begin{aligned} B' &= \sum_{i=0}^{e-1}\sum_{j=0}^{z-1} F\cdot\Omega(\lambda v_\theta)^i v_\phi^j \\ &= [\sum_{i=0}^{e-1} F(\lambda v_\theta)^i]\cdot[\sum_{j=0}^{z-1} \Omega v_\phi^j] \\ &\approx ((\lambda v_\theta)^e,\ F/k,\ \theta) \otimes_k (v_\phi^z,\ \Omega/k,\ \phi) \\ &\sim (\lambda^e v_\theta^e,\ F/k,\ \theta). \end{aligned}$$

Here, $(v_\phi^z,\ \Omega/k,\ \phi) \sim k$, because $v_\phi^z = \alpha'(\phi,\phi)\alpha'(\phi^2,\phi)\cdots\alpha'(\phi^{z-1},\phi)$ is a root of unity and Ω/k is unramified. We have

$$\begin{aligned} \delta &= \lambda^e v_\theta^e = \lambda^e\alpha'(\theta,\theta)\alpha'(\theta^2,\theta)\cdots\alpha'(\theta^{e-1},\theta) \\ &= (\alpha(\omega,\eta)/\alpha(\eta,\omega))^{e/(q-1)}\alpha(\omega,\omega)\alpha(\omega^2,\omega)\cdots\alpha(\omega^{e-1},\omega). \end{aligned}$$

As δ is a root of unity in k whose order is relatively prime to p, δ is expressed as $\delta = \zeta_{q-1}^{v}$ for a certain integer v.

Now we are going to calculate the index of the above cyclic algebra

$$(\delta,\ F/k,\ \theta).$$

It is the smallest positive integer m such that δ^m is in the norm group $N_{F/k}(F^*) = \{N_{F/k}(x);\ x \in F^*\}$. Recall that F is a totally ramified extension of k of degree e, and that e is also the ramification index of $L' = Q_p(\zeta_{q^z-1}, \zeta_{p^n})$ over k. Since $e_{k/Q_p} = cp^d$, $(c, p) = 1$, it follows that

$$e = p^{n-1-d}(p-1)/c.$$

Let π be a prime element of F. It is wellknown that every element x of F^* has the unique expression:

$$x = \pi^s \zeta_{q-1}^r \rho,$$

$$s \in Z,\quad r \bmod q-1,\quad \rho\text{: principal unit of } F.$$

Note that

$$N_{F/k}(<\zeta_{q-1}>) = <\zeta_{q-1}^{e}> = <\zeta_{q-1}^{(p-1)/c}>.$$

Hence we have

$$N_{F/k}(F^*) = \{N_{F/k}(\pi)^s \zeta_{q-1}^{r(p-1)/c} N_{F/k}(\rho);$$

$$s \in Z,\quad r \bmod ((q-1)/\tfrac{p-1}{c}),\quad \rho\text{: principal unit of } F\}$$

Note that $N_{F/k}(\rho)$ is a principal unit of k. Let m be an integer. Then, $\delta^m = \zeta_{q-1}^{vm}$ is in $N_{F/k}(F^*)$ if and only if there exist some s, r, and ρ such that

$$\zeta_{q-1}^{vm} = N_{F/k}(\pi)^s \zeta_{q-1}^{r(p-1)/c} N_{F/k}(\rho).$$

But this holds if and only if

$$s = 0, \quad N_{F/k}(\rho) = 1, \quad vm \equiv r(p-1)/c \pmod{q-1}.$$

Hence the smallest positive integer m such that δ^m is in $N_{F/k}(F^*)$ is equal to

$$\frac{(p-1)/c}{(v, (p-1)/c)} .$$

Next we will prove $(\gamma, L/k) \sim k$. Recall that $\gamma(\sigma, \tau) \in \mathcal{W}_p(L) = \langle \zeta_{p^n} \rangle$. Denote by $\gamma'(\iota, \kappa) = (\mathrm{Cor}\gamma)(\iota, \kappa)$, $\iota, \kappa \in G(L/Q_p)$, the image of the 2-cocycle $\gamma(\sigma, \tau)$ by the corestriction homomorphism : $H^2(L/k) \to H^2(L/Q_p)$. It is wellknown that this homomorphism is injective. Hence the order of the 2-cocycle γ in $H^2(L/k)$ is equal to that of the 2-cocycle γ' in $H^2(L/Q_p)$. It follows from the definition of corestriction homomorphism that $\gamma'(\iota, \kappa) \in \mathcal{W}_p(L) = \langle \zeta_{p^n} \rangle$. We have only to prove that the following crossed product with the factor set γ' is similar to Q_p:

$$(\gamma'(\iota, \kappa), L/Q_p) = \sum_{\iota \in G'} L v_\iota, \quad L = Q_p(\zeta_{p^n}, \zeta_{q^f-1}),$$

$$v_\iota v_\kappa = \gamma'(\iota, \kappa) v_{\iota\kappa}, \quad v_\iota x v_\iota^{-1} = x^\iota \ (x \in L),$$

where $G' = G(L/Q_p)$. Put $F = Q_p(\zeta_{p^n})$, $\Omega = Q_p(\zeta_{q^f-1})$. Then F/Q_p is totally ramified, Ω/Q_p is unramified, $F\cdot\Omega = L$, and $F\cap\Omega = Q_p$. We have

$$G(L/Q_p) = \langle\theta\rangle \times \langle\phi\rangle,$$

$$G(F/Q_p) \simeq G(L/\Omega) = \langle\theta\rangle, \quad G(\Omega/Q_p) \simeq G(L/F) = \langle\phi\rangle,$$

$$\zeta_{p^n}^{\theta} = \zeta_{p^n}^{r}, \; \zeta_{q^f-1}^{\theta} = \zeta_{q^f-1}, \; \zeta_{p^n}^{\phi} = \zeta_{p^n}, \; \zeta_{q^f-1}^{\phi} = \zeta_{q^f-1}^{p},$$

where r is a primitive root modulo p^n, and ϕ is a Frobenius automorphism of L/Q_p. It is easy to see that

$$(p, r-1) = 1 \quad\text{and}\quad \langle\zeta_{p^n}^{r-1}\rangle = \langle\zeta_{p^n}\rangle.$$

Hence we can write

$$\gamma'(\phi, \theta)/\gamma'(\theta, \phi) = \rho^{r-1} \quad\text{for some}\quad \rho \in \langle\zeta_{p^n}\rangle.$$

Then we have

$$v_\phi v_\theta = (\gamma'(\phi,\theta)/\gamma'(\theta,\phi))v_\theta v_\phi = \rho^{r-1} v_\theta v_\phi,$$

$$(\rho v_\phi)v_\theta = \rho^{r} v_\theta v_\phi = v_\theta(\rho v_\phi).$$

Put $e = p^{n-1}(p-1) = |\langle\theta\rangle|$, $f' = |\langle\phi\rangle|$. Since ρv_ϕ commutes with v_θ, and ρv_ϕ (resp. v_θ) commutes with each element of F (resp. Ω), it follows that

$$(\gamma', L/Q_p) = \sum_{\iota\in G'} L v_\iota = \sum_{i=0}^{e-1}\sum_{j=0}^{f'-1} F\cdot\Omega v_\theta^{i}(\rho v_\phi)^{j}$$

$$= (\sum_{i=0}^{e-1} Fv_\theta^i)\cdot(\sum_{j=0}^{f'-1} \Omega(\rho v_\phi)^j)$$

$$= (\lambda, F/Q_p, \theta) \otimes_{Q_p} (\lambda', \Omega/Q_p, \phi),$$

$$\sim (\lambda, F/Q_p, \theta),$$

where

$$\lambda = v_\theta^e = \gamma'(\theta, \theta)\gamma'(\theta^2, \theta)\cdots\gamma'(\theta^{e-1}, \theta),$$

$$\lambda' = (\rho v_\phi)^{f'} = \rho^{f'} v_\phi^{f'} = \rho^{f'}\gamma'(\phi,\phi)\gamma'(\phi^2,\phi)\cdots\gamma'(\phi^{f'-1},\phi),$$

and Ω/Q_p is unramified. Since λ belongs to $\langle\zeta_{p^n}\rangle \cap Q_p = \{1\}$, we conclude that $\lambda = 1$ and $(\lambda, F/Q_p, \theta) \sim Q_p$. This completes the proof of Theorem 4.3.

Using Theorem 4.3, we can determine the Schur subgroup $S(k)$.

Theorem 4.4 (Yamada [45]). Let p be an odd prime. Let k be a cyclotomic extension of Q_p. Let c be the tame ramification index of k/Q_p. Then an algebra class $[A]$ of $Br(k)$ belongs to $S(k)$ if and only if the invariant of A is of the form:

$$\mathrm{inv}_k(A) \equiv z/\tfrac{p-1}{c} \pmod{Z} \quad \text{for some} \quad z \in Z.$$

Proof. By Corollary 3.11, $S(k)$ consists of those algebra classes represented by a cyclotomic algebra over k. Hence "only if" part of the theorem follows at once from Theorem 4.3.

To prove "if" part of the theorem, we need only show that

for every $z \in Z$, there exists a cyclotomic algebra over k whose Hasse invariant is $z/\frac{p-1}{c}$. Clearly it suffices to show the existence of a cyclotomic algebra over k, whose index is $(p-1)/c$. Let $L = Q_p(\zeta)$ be a cyclotomic field containing k, ζ being some root of unity. Put $e = e_{L/k}$, $f = f_{L/k}$, $z = ef$. Put $\Omega = k(\zeta_{q^z-1})$, where $q = p^{f^*}$, $f^* = f_{k/Q_p}$. Ω is the unramified extension of k of degree z. Put $L' = L\cdot\Omega$. It is easy to see that $L' = Q_p(\zeta_{p^n}, \zeta_{q^z-1})$ for some integer n. We may assume $n \geqq 1$. Lemma 4.1 yields that there exists a subfield F of L' over k, such that F is totally ramified over k, $[F : k] = e$, $F\cdot\Omega = L'$, and $F \cap \Omega = k$. Let

$$G(L'/k) = \langle\omega\rangle \times \langle\eta\rangle, \quad G(L'/\Omega) = \langle\omega\rangle, \quad G(L'/F) = \langle\eta\rangle,$$

η being a Frobenius automorphism of L'/k, so that $\zeta_{q^z-1}^{\eta} = \zeta_{q^z-1}^{q}$. Put

$$h_{\omega,\eta} = h_{\eta,\omega} = 1, \quad h_\omega = \zeta_{q-1}, \quad h_\eta = 1.$$

It is easy to see that these h's satisfy the relations (2.16)-(2.18) and hence give rise to a cyclotomic algebra over k:

$$B = (\alpha, L'/k) = \sum_{i=0}^{e-1}\sum_{j=0}^{z-1} L'u_\omega^i u_\eta^j,$$

$$u_\omega^i u_\eta^j x = x^{\omega^i\eta^j} u_\omega^i u_\eta^j \quad (x \in L'),$$

$$u_\omega u_\eta = u_\eta u_\omega, \quad u_\omega^e = \zeta_{q-1}, \quad u_\eta^z = 1,$$

$$(u_\omega^i u_\eta^j)(u_\omega^{i'} u_\eta^{j'}) = \alpha(\omega^i\eta^j, \omega^{i'}\eta^{j'})u_\omega^{i''}u_\eta^{j''},$$

$$0 \leq i, i', i'' \leq e - 1, \quad 0 \leq j, j', j'' \leq z - 1,$$

$$i'' \equiv i + i' \pmod{e}, \quad j'' \equiv j + j' \pmod{z}.$$

(See chapter 2.) We have

$$\alpha(\omega, \eta)/\alpha(\eta, \omega) = u_\omega u_\eta u_\omega^{-1} u_\eta^{-1} = 1,$$

$$\alpha(\omega, \omega)\alpha(\omega^2, \omega)\cdots\alpha(\omega^{e-1}, \omega) = u_\omega^e = \zeta_{q-1},$$

$$\delta = 1^{e/(q-1)}\cdot\zeta_{q-1} = \zeta_{q-1}^{1+e}.$$

We notice that $e \equiv 0 \pmod{(p-1)/c}$, and that

$$\frac{(p - 1)/c}{(1 + e, (p - 1)/c)} = (p - 1)/c.$$

Consequently, Theorem 4.3 yields that the index of the cyclotomic algebra B over k is equal to $(p - 1)/c$, establishing Theorem 4.4.

It is obvious that Theorem 4.4 is equivalent to the following

<u>Theorem 4.4'</u>. Let p be an odd prime and k a cyclotomic extension of Q_p. Let $Q_p(\zeta)$ be any arbitrary cyclotomic field containing k and ζ_p, a primitive p-th root of unity. Let b be the tame ramification index of $Q_p(\zeta)/k$. Then $S(k)$ is the subgroup of $Br(k)$ of order b.

<u>Remark 4.5</u>. $b = (p - 1)/c$, where c is as in Theorem 4.4.

Finally we state

Proposition 4.6. Let K be a finite extension of Q_p, p being an arbitrary prime number. Let k denote the maximal cyclotomic extension of Q_p contained in K. Then

$$S(K) = S(k) \otimes_k K = \{[A \otimes K];\ [A] \in S(k)\}.$$

Proof. Let B be a cyclotomic algebra over K: $B = (\beta, K(\zeta)/K)$. We may assume $\beta(\sigma, \tau) \in \langle\zeta\rangle$ for all $\sigma, \tau \in G(K(\zeta)/K)$. The intersection $K \cap k(\zeta)$ is a cyclotomic extension of Q_p contained in K, so that $K \cap k(\zeta) = k$. Then $G(K(\zeta)/K)$ is canonically isomorphic to $G(k(\zeta)/k)$, and $(\beta, K(\zeta)/K) \simeq (\beta, k(\zeta)/k) \otimes_k K$. The proposition now follows immediately.

Theorem 4.7 (Yamada [45]). Let K be a finite extension of Q_p, p being an odd prime. Let k denote the maximal cyclotomic extension of Q_p contained in K. Let c be the tame ramification index of k/Q_p. Put $s = ((p - 1)/c, [K : k])$. Then

$$S(K) = \{[B] \in Br(K);\ \mathrm{inv}_K(B) \equiv \frac{z}{(p-1)/(cs)} \pmod{Z},\quad z \in Z\}.$$

Proof. This follows at once from Theorem 4.4 and Proposition 4.6.

Proposition 4.8. Let K be an extension of Q_p. If a primitive p-th root of unity ζ_p belongs to K, then $S(K) = 1$.

<u>Proof</u>. Let A be a Schur algebra over K. There exists a finite group G and its irreducible character χ such that $K(\chi) = K$ and $A = A(\chi, K)$. Put $k = Q_p(\chi, \zeta_p)$. Since $K \supset k \supset Q_p(\chi)$, we have $A = A(\chi, k) \otimes_k K$. Because $\zeta_p \in k$ and so the tame ramification index of k/Q_p equals $p - 1$, it follows from Theorem 4.4 that $S(k) = 1$, whence $A(\chi, k) \sim k$ and $A \sim K$. Thus we conclude $S(K) = 1$. #

Appendix to Chapter 4.

(Determination of local indices of a cyclotomic algebra over a number field.)

Let k be a cyclotomic extension of Q, the rationals. Let p be a prime number and $\mathfrak{p}$ a prime of k lying above p. Let

$$B = (\beta, Q(\zeta)/k) = \sum_{\sigma \in G} Q(\zeta)u_\sigma, \quad (u_1 = 1), \quad G = G(Q(\zeta)/k) \qquad (4.10)$$

be a cyclotomic algebra over k, where ζ is a root of unity such that $Q(\zeta) \supset k$. Then $G_{\mathfrak{p}} = G(Q(\zeta)^{\mathfrak{p}}/k_{\mathfrak{p}})$ can be regarded as a subgroup of $G = G(Q(\zeta)/k)$, the decomposition group of $\mathfrak{p}$ in $Q(\zeta)/k$, and

$$B_{\mathfrak{p}} = B \otimes_k k_{\mathfrak{p}} \sim (\beta_{\mathfrak{p}}, Q(\zeta)^{\mathfrak{p}}/k_{\mathfrak{p}}) = \sum_{\tau \in G_{\mathfrak{p}}} Q(\zeta)^{\mathfrak{p}} u_\tau, \qquad (4.11)$$

where $\beta_{\mathfrak{p}}$ denotes the restriction of β to $G_{\mathfrak{p}}$. The right side is a cyclotomic algebra over $k_{\mathfrak{p}}$ $(\supset Q_p)$. Hence if p is odd, we can use the formula (4.9) to determine the $\mathfrak{p}$-local index of B. The procedure is as follows: We may assume that $Q(\zeta) \ni \zeta_p$ and that the order of $\beta(\sigma, \tau)$ is relatively prime to p for any $\sigma, \tau \in G$. Find a generator ω of the inertia group $T_{\mathfrak{p}}$ and a Frobenius automorphism η of $\mathfrak{p}$ in $Q(\zeta)/k$. Determine the ramification index e of $\mathfrak{p}$ in $Q(\zeta)/k$, the tame ramification index c of p in k/Q, and $q = N_{k/Q}(\mathfrak{p})$. Compute the number δ in Theorem 4.3 and find an integer v such that $\delta = \zeta_{q-1}^{v}$. Then, use the formula (4.9) to know the p-local index of B.

If $G_p = G(Q(\zeta)^p/k_p)$ is cyclic, $G_p = \langle\xi\rangle$, and $|\langle\xi\rangle| = t$, then B_p is similar to a cyclic algebra over k_p:

$$B_p \sim (\beta', Q(\zeta)^p/k_p) = \sum_{i=0}^{t-1} Q(\zeta)^p u_\xi^i,$$

$$\beta'(\xi^i, \xi^j) = \begin{cases} u_\xi^t, & \text{if } i + j \geqq t, \\ 1, & \text{if } i + j < t, \end{cases}$$

where $0 \leqq i, j \leqq t - 1$. Recall that the factor set of a cyclic algebra is "commutative". That is, $\beta'(\xi^i, \xi^j) = \beta'(\xi^j, \xi^i)$. Hence, for $\langle\omega\rangle = T_p$ and a Frobenius automorphism η of p in $Q(\zeta)/k$, ($\omega, \eta \in \langle\xi\rangle = G_p$), we have

$$(\beta'(\omega, \eta)/\beta'(\eta, \omega))^{e/(q-1)} = 1^{e/(q-1)} = \zeta_{q-1}^{e}. \tag{4.12}$$

Let ν be the least positive integer such that $\xi^\nu \in T_p$. Then $\nu e = t$ and $\langle\xi^\nu\rangle = T_p$, so it may be assumed that $\omega = \xi^\nu$. Then we see that

$$\beta'(\omega, \omega)\beta'(\omega^2, \omega)\cdots\beta'(\omega^{e-1}, \omega) = u_\xi^t$$

$$= \beta(\xi, \xi)\beta(\xi^2, \xi)\cdots\beta(\xi^{t-1}, \xi) \in W'(k_p) \subset k_p. \tag{4.13}$$

Hence we can write

$$u_{\xi}^{t} = \zeta_{q-1}^{v'} \qquad (4.14)$$

for some integer v'. Then by (4.12) and (4.14), the number δ in Theorem 4.3 is: $\delta = \zeta_{q-1}^{v}$, $v = e + v'$. Note that e is divisible by $(p - 1)/c$, because $\zeta_p \in Q(\zeta)$. By the formula (4.9) of Theorem 4.3, we thus conclude that the p-local index of B is equal to

$$\frac{(p - 1)/c}{(v, (p - 1)/c)} = \frac{(p - 1)/c}{(v', (p - 1)/c)} . \qquad (4.15)$$

In particular, if B itself is a cyclic algebra over k:

$$B = (a, Q(\zeta)/k, \sigma) = \sum_{i=0}^{s-1} Q(\zeta)u_{\sigma}^{i}, \qquad u_{\sigma}^{s} = a,$$

$$\langle\sigma\rangle = G = G(Q(\zeta)/k), \qquad |\langle\sigma\rangle| = s,$$

then

$$B_p = B \otimes_k k_p \sim (a, Q(\zeta)^p/k_p, \sigma^{\lambda}),$$

where λ is the least positive integer such that $\sigma^{\lambda} \in G_p = G(Q(\zeta)^p/k_p)$, and so $\langle\sigma^{\lambda}\rangle = G_p$. We see that the quantities σ^{λ}, $\frac{s}{\lambda}$, u_{σ}^{λ} correspond to the previous quantities ξ, t, u_{ξ}, respectively. Therefore, when we write

$$a = u_\sigma^s = (u_\sigma^\lambda)^{s/\lambda} = \zeta_{q-1}^{v'}, \quad v' \in Z, \tag{4.16}$$

then the p-index of B is equal to

$$\frac{(p-1)/c}{(v', (p-1)/c)} . \tag{4.17}$$

Under a certain condition we can also simplify the process for determining the p-index of $B = (\beta, Q(\zeta)/k)$, even though $G(Q(\zeta)^p/k_p)$ is not cyclic. Let

$$G = G(Q(\zeta)/k) = \langle\phi_1\rangle \times \langle\phi_2\rangle \times \cdots \times \langle\phi_s\rangle, \tag{4.18}$$

where $\langle\phi_i\rangle$ is a cyclic group of order n_i. Then

$$B = (\beta, Q(\zeta)/k) = (\beta', Q(\zeta)/k)$$

$$= \sum_{i_1=0}^{n_1-1} \cdots \sum_{i_s=0}^{n_s-1} Q(\zeta) u_{\phi_1}^{i_1} \cdots u_{\phi_s}^{i_s}, \tag{4.19}$$

$$(u_{\phi_1}^{i_1} \cdots u_{\phi_s}^{i_s})(u_{\phi_1}^{i'_1} \cdots u_{\phi_s}^{i'_s})$$

$$= \beta'(\phi_1^{i_1} \cdots \phi_s^{i_s}, \phi_1^{i'_1} \cdots \phi_s^{i'_s})\, u_{\phi_1}^{i''_1} \cdots u_{\phi_s}^{i''_s}, \tag{4.20}$$

$$0 \leqq i_\nu, i'_\nu, i''_\nu \leqq n_\nu - 1, \quad i''_\nu \equiv i_\nu + i'_\nu \pmod{n_\nu}.$$

$(\nu = 1, 2, \cdots, s)$. To determine the p-index of B, we may also use the number

$$\delta' = (\beta'(\omega,\eta)/\beta'(\eta,\omega))^{e/(q-1)}\beta'(\omega,\omega)\beta'(\omega^2,\omega)\cdots\beta'(\omega^{e-1},\omega).$$

Write

$$\beta'(\omega,\omega)\beta'(\omega^2,\omega)\cdots\beta'(\omega^{e-1},\omega) = \zeta_{q-1}^{v'}, \quad (v' \in Z) \tag{4.21}$$

$$\omega = \prod_{i=1}^{s} \phi_i^{a_i}, \qquad \eta = \prod_{i=1}^{s} \phi_i^{b_i}, \tag{4.22}$$

$$0 \leqq a_i, b_i \leqq n_i - 1 \quad (i = 1, 2, \cdots, s), \tag{4.23}$$

$$v_\omega = u_{\phi_1}^{a_1}\cdots u_{\phi_s}^{a_s}, \qquad v_\eta = u_{\phi_1}^{b_1}\cdots u_{\phi_s}^{b_s}. \tag{4.24}$$

Then

$$v_\omega v_\eta = (\beta'(\omega, \eta)/\beta'(\eta, \omega))v_\eta v_\omega. \tag{4.25}$$

If it happens that

$$v_\omega v_\eta = v_\eta v_\omega \quad \text{or} \quad \beta'(\omega, \eta)/\beta'(\eta, \omega) = 1, \tag{4.26}$$

then $\delta' = 1^{e/(q-1)}\cdot\zeta_{q-1}^{v'} = \zeta_{q-1}^{v}$, $v = e + v'$. Since $\zeta_p \in Q(\zeta)$, it follows that e is divisible by $(p - 1)/c$. Hence, by the formula (4.9) of Theorem 4.3 we conclude that the p-index of B is equal to $\{(p - 1)/c\}/(v', (p - 1)/c)$.

In many cases, we can choose direct factors $\langle\phi_i\rangle$ of G so that the inertia group $\langle\omega\rangle = T_p$ is a subgroup of one of factors, say $\langle\phi_1\rangle : \langle\omega\rangle \subset \langle\phi_1\rangle$. Let d be the least positive integer such that $\phi_1^d \in \langle\omega\rangle$. Then $de = n_1$, and we may assume $\phi_1^d = \omega$. It follows from the definition of the factor set β' that

$$\beta'(\omega^i, \omega) = 1 \quad (1 \leq i \leq e - 2),$$

$$\beta'(\omega^{e-1}, \omega) = u_{\phi_1}^{n_1} = \beta(\phi_1, \phi_1)\beta(\phi_1^2, \phi_1)\cdots\beta(\phi_1^{n_1-1}, \phi_1),$$

$$u_{\phi_1}^{n_1} = \beta'(\omega, \omega)\beta'(\omega^2, \omega)\cdots\beta'(\omega^{e-1}, \omega) = \zeta_{q-1}^{v'}.$$

Hence we have proved:

Proposition 4.9. Let the notation be as before. If $v_\omega v_\eta = v_\eta v_\omega$ (i.e., $\beta'(\omega, \eta)/\beta'(\eta, \omega) = 1$), $\omega = \phi_1^d$, $d | n_1$, and $u_{\phi_1}^{n_1} = \zeta_{q-1}^{v'}$, then the p-index of B is equal to

$$\frac{(p - 1)/c}{(v', (p - 1)/c)} . \tag{4.27}$$

Chapter 5. THE SCHUR SUBGROUP OF A 2-ADIC FIELD

In this chapter, k is a cyclotomic extension of Q_2, the rational 2-adic numbers. Let B be an arbitrary cyclotomic algebra over k. It was shown at the beginning of Chapter 4 that B may be assumed to be of the form:

$$B = (\beta, L/k) = \sum_{\sigma\varepsilon G} Lu_\sigma, \quad (u_1 = 1), \tag{5.1}$$

$$L = Q_p(\zeta_{2^n}, \zeta_{q^f-1}), \quad q = 2^{f^*}, \quad f^* = f_{k/Q_2}, \quad f = f_{L/k}, \tag{5.2}$$

$$u_\sigma u_\tau = \beta(\sigma, \tau)u_{\sigma\tau}, \quad u_\sigma x = x^\sigma u_\sigma \quad (x \in L), \tag{5.3}$$

$$\beta(\sigma,\tau) = \alpha(\sigma,\tau)\gamma(\sigma,\tau), \quad \alpha(\sigma,\tau) \in \langle\zeta_{q^f-1}\rangle, \quad \gamma(\sigma,\tau) \in \langle\zeta_{2^n}\rangle. \tag{5.4}$$

$$(\beta, L/k) \sim (\alpha, L/k) \otimes_k (\gamma, L/k), \tag{5.5}$$

where $\sigma, \tau \in G = G(L/k)$. We will see that $(\alpha, L/k) \sim k$ (cf. Proposition 5.1). For simplicity, put

$$r = q^f - 1 = 2^{f^*f} - 1. \tag{5.6}$$

If $n \leq 1$, then $B \sim k$, because the extension L/k is unramified and the factor set β consists of roots of unity. Hence we always assume $n \geq 2$.

Let T_0 denote the inertia group of L/Q_2. Then

$$T_0 = \langle\theta\rangle \times \langle\iota\rangle, \quad \theta^{2^{n-2}} = \iota^2 = 1, \tag{5.7}$$

$$\zeta_{2^n}^{\theta} = \zeta_{2^n}^{5}, \quad \zeta_{2^n}^{\iota} = \zeta_{2^n}^{-1}, \quad \zeta_r^{\theta} = \zeta_r^{\iota} = \zeta_r. \tag{5.8}$$

A Frobenius automorphism ξ of L/Q_2 is given by

$$\zeta_r^{\xi} = \zeta_r^{2}, \quad \zeta_{2^n}^{\xi} = \zeta_{2^n}. \tag{5.9}$$

We have

$$G(L/Q_2) = \langle\theta\rangle \times \langle\iota\rangle \times \langle\xi\rangle. \tag{5.10}$$

The subgroups of T_0 are classified into three types:

(i) $\langle\theta^{2^\lambda}\rangle \times \langle\iota\rangle$, $(\lambda = 0,1,\cdots,n-2)$,

(ii) $\langle\theta^{2^\lambda}\rangle$, $(\lambda = 0,1,\cdots,n-2)$,

(iii) $\langle\iota\theta^{2^\nu}\rangle$, $(\nu = 0,1,\cdots,n-3)$.

(Type (i) is the "non-cyclic" case. If $n = 2$, then $\theta = 1$, $T_0 = \langle\iota\rangle$, and hence the type (iii) does not arise.) Let T denote the inertia group of L/k. Then $T = T_0 \cap G(L/k)$, so T belongs to one of the above three types.

We apply Lemma 4.1 to the extension L/k. Put

$$e = e_{L/k}, \quad f = f_{L/k}, \quad z = ef, \quad f^* = f_{k/Q_2}. \tag{5.11}$$

Denote by Ω the unramified extension of k of degree z and set $L' = L\cdot\Omega$. Then

$$L' = Q_2(\zeta_{2^n}, \zeta_{r'}), \quad r' = 2^{f^*z} - 1, \quad z = f_{L'/k}. \tag{5.12}$$

It follows from Lemma 4.1 that there exists a totally ramified extension F of k of degree e such that $F \cdot \Omega = L'$, $F \cap \Omega = k$, and $G(L'/\Omega)$ is canonically isomorphic to T, the inertia group of L/k. We can describe the circumstances more explicitly. We may obviously write

$$G(L'/Q_2) = \langle\theta\rangle \times \langle\iota\rangle \times \langle\xi'\rangle, \tag{5.13}$$

where θ and ι are defined by (5.8) with modification of $\zeta_{r'}^{\theta} = \zeta_{r'}^{\iota} = \zeta_{r'}$, and where ξ' is defined by $\zeta_{r'}^{\xi'} = \zeta_{r'}^{2}$, $\zeta_{2^n}^{\xi'} = \zeta_{2^n}$. Let T' denote the inertia group of L'/k. Then we can identify T' with T. Namely, if $T = \langle\theta^{2^\lambda}\rangle \times \langle\iota\rangle \subset G(L/k)$, then $T' = \langle\theta^{2^\lambda}\rangle \times \langle\iota\rangle \subset G(L'/k)$. Also, if $T = \langle\theta^{2^\lambda}\rangle$ (resp. $T = \langle\iota\theta^{2^\nu}\rangle) \subset G(L/k)$, then $T' = \langle\theta^{2^\lambda}\rangle$ (resp. $T' = \langle\iota\theta^{2^\nu}\rangle) \subset G(L'/k)$. Let η be a Frobenius automorphism of L/k. Regarding η as an automorphism of L/Q_2, we can write

$$\eta = \xi^{f^*}\theta^{x}\iota^{y} \qquad (1 \leq x \leq 2^{n-2};\ \ y = 0, 1) \tag{5.14}$$

for some integers x, y, $(f^* = f_{k/Q_2})$. Since $[L : k] = ef = z$, we have $\eta^z = 1$, i.e., $\xi^{f^*z} = \theta^{xz} = \iota^{yz} = 1$. Put

$$\phi = (\xi')^{f^*}\theta^{x}\iota^{y} \in G(L'/k). \tag{5.15}$$

It is easy to see that ϕ is a Frobenius automorphism of L'/k, $\phi \mid L = \eta$, i.e., $\phi(\delta) = \eta(\delta)$ for any $\delta \in L$, and $\phi^z = 1$, $(z = f_{L'/k})$. Hence

$$G(L'/k) = T' \times \langle\phi\rangle, \quad (\phi^z = 1). \tag{5.16}$$

Let Inf denote the inflation map from $H^2(L/k)$ into $H^2(L'/k)$ and put $\beta' = \mathrm{Inf}(\beta)$. Then

$$B \sim B' = (\beta', L'/k) = \sum_{\sigma\in G'} L'v_\sigma, \tag{5.17}$$

$$v_\sigma v_\tau = \beta'(\sigma, \tau)v_{\sigma\tau}, \quad v_\sigma\lambda = \lambda^\sigma v_\sigma \quad (\lambda \in L'),$$

where $\sigma, \tau \in G' = G(L'/k)$. It follows from the definition of inflation map that

$$\beta'(\tau\phi^\nu, \tau'\phi^{\nu'}) = \beta(\tau\eta^\nu, \tau'\eta^{\nu'}) \tag{5.18}$$

for any $\tau, \tau' \in T' \simeq T$ and ν, ν' $(0 \leqq \nu, \nu' \leqq z - 1)$. Hence we have

$$v_\tau v_\phi v_\tau^{-1} v_\phi^{-1} = \beta'(\tau,\phi)/\beta'(\phi,\tau) = \beta(\tau,\eta)/\beta(\eta,\tau) = u_\tau u_\eta u_\tau^{-1} u_\eta^{-1}, \tag{5.19}$$

$$v_\tau^d = \beta(\tau,\tau)\beta(\tau^2,\tau)\cdots\beta(\tau^{d-1},\tau) = u_\tau^d, \tag{5.20}$$

$(u_1 = v_1 = 1;$ d being the order of $\tau)$,

$$v_\phi^z = \beta'(\phi,\phi)\beta'(\phi^2,\phi)\cdots\beta'(\phi^{z-1},\phi)$$
$$= \beta(\eta,\eta)\beta(\eta^2,\eta)\cdots\beta(\eta^{z-1},\eta). \tag{5.21}$$

Here we note that for an element σ of $G(L'/k)$ whose order is h, v_σ^h is a root of unity in $L \subset L'$ and $(v_\sigma^h)^\sigma = v_\sigma^h$, because $\beta' = \mathrm{Inf}(\beta)$, $v_\sigma^h = \beta'(\sigma,\sigma)\beta'(\sigma^2,\sigma)\cdots\beta'(\sigma^{h-1},\sigma)$, and $(v_\sigma^h)^\sigma = v_\sigma v_\sigma^h v_\sigma^{-1} = v_\sigma^h$. Let Ω (resp. F) be the subfield of L' over k corresponding to T' (resp. $\langle\phi\rangle$) in the sense of Galois theory.

Then $\Omega\cdot F = L'$, $\Omega \cap F = k$, $T' = G(L'/\Omega) \simeq G(F/k)$, and $\langle\phi\rangle = G(L'/F) \simeq G(\Omega/k)$. It follows from (5.17) that

$$B = (\beta, L/k) \sim B' = (\beta', L'/k)$$

$$= \sum_{\tau\in T'} \sum_{\nu=0}^{z-1} \Omega\cdot F v_\tau v_\phi^\nu. \qquad (5.22)$$

For the rest of this chapter, we will use the same notation as in (5.1)-(5.22).

<u>Proposition 5.1</u>. Let $B = (\beta, L/k)$ be a cyclotomic algebra over k defined by (5.1)-(5.6), and such that $\beta(\sigma, \tau) \in \langle\zeta_r\rangle = w'(L)$. Then, $B \sim k$.

<u>Proof</u>. Let the notation be as in (5.7)-(5.22). Recall that $w'(L') = \langle\zeta_{q'-1}\rangle$, $w'(k) = \langle\zeta_{q-1}\rangle$, where $q = 2^{f^*}$, $q' = q^z$. Suppose first that the inertia group T of L/k is "cyclic", i.e., $T = \langle\tau\rangle$, where either $\tau = \theta^{2^\lambda}$ or $\iota\theta^{2^\nu}$. We note that $\zeta_{q'-1}^\phi = \zeta_{q'-1}^q$, $\zeta_{q'-1}^\tau = \zeta_{q'-1}$. By (5.19), (5.21), (5.22) we have

$$B \sim B' = (\beta', L'/k) = \sum_{i=0}^{e-1} \sum_{j=0}^{z-1} \Omega\cdot F v_\tau^i v_\phi^j,$$

$$v_\tau v_\phi = \zeta_{q'-1}^b v_\phi v_\tau, \quad v_\phi^z = \zeta_{q-1}^\mu, \quad (\zeta_{q-1} = \zeta_{q'-1}^{(q'-1)/(q-1)})$$

for some integers b, μ. From the relation (2.18) it follows that

$$1 = (\zeta_{q-1}^\mu)^{\tau-1} = (\zeta_{q'-1}^b)^{1+\phi+\cdots+\phi^{z-1}}$$

$$= (\zeta_{q'-1}^{b})^{1+q+\cdots+q^{z-1}} = \zeta_{q'-1}^{b(q'-1)/(q-1)} = \zeta_{q-1}^{b}. \qquad (5.23)$$

Hence $(q-1)|b$, and so there exists an integer X satisfying $(q - 1)X \equiv b \pmod{q'-1}$ Set $w_\tau = \zeta_{q'-1}^{X} v_\tau$. Then

$$v_\phi w_\tau = \zeta_{q'-1}^{qX-b} v_\tau v_\phi = \zeta_{q'-1}^{X} v_\tau v_\phi = w_\tau v_\phi,$$

so we have

$$B' = \sum_{i=0}^{e-1} \sum_{j=0}^{z-1} \Omega \cdot F w_\tau^i v_\phi^j$$

$$= (w_\tau^e, F/k, \tau) \otimes_k (v_\phi^z, \Omega/k, \phi),$$

$$w_\tau^e = \zeta_{q'-1}^{eX} v_\tau^e, \quad v_\phi^z \in \mathcal{W}'(k) = \langle \zeta_{q-1} \rangle.$$

Note that $\mathcal{W}'(Q_2) = \{1\}$ and $N_{k/Q_2}(\zeta_{q-1}) \in \mathcal{W}'(Q_2)$, so $N_{k/Q_2}(\zeta_{q-1}) = 1$. Since the index of the cyclic algebra $(w_\tau^e, F/k, \tau)$ is equal to the order of the norm residue symbol $(w_\tau^e, F/k) = (N_{k/Q_2}(w_\tau^e), F/Q_2) = (1, F/Q_2) = 1 \in G(F/Q_2)$ it follows that $(w_\tau^e, F/k, \tau) \sim k$. Similarly, $(v_\phi^{f'}, \Omega/k, \phi) \sim k$, and hence $B \sim B' \sim k$.

Suppose next that $T = \langle\tau\rangle \times \langle\iota\rangle$, $\tau = \theta^{2^\lambda}$. Then $G(L'/k) = \langle\tau\rangle \times \langle\iota\rangle \times \langle\phi\rangle$, $\tau^{e/2} = \iota^2 = \phi^z = 1$. Let Ω, F_1, F_2 be the subfields of L' over k corresponding to $\langle\tau\rangle \times \langle\iota\rangle$, $\langle\iota\rangle \times \langle\phi\rangle$, $\langle\tau\rangle \times \langle\phi\rangle$ respectively in the sense of Galois theory. Then $G(\Omega/k) \simeq \langle\phi\rangle$, $G(F_1/k) \simeq \langle\tau\rangle$, $G(F_2/k) \simeq \langle\iota\rangle$. It follows from (5.19)-(5.22) that

$$B \sim B' = \sum_{i=0}^{e/2-1} \sum_{j=0}^{1} \sum_{\nu=0}^{z-1} \Omega \cdot F_1 \cdot F_2 \; v_\tau^i v_\iota^j v_\phi^\nu, \tag{5.24}$$

$$v_\tau v_\phi = \zeta_{q'-1}^{b} v_\phi v_\tau, \quad v_\iota v_\phi = \zeta_{q'-1}^{b'} v_\phi v_\iota, \quad v_\iota v_\tau = \zeta_{q'-1}^{b''} v_\tau v_\iota,$$

$$v_\phi^z = \zeta_{q-1}^{\mu}, \quad v_\iota^2 = \zeta_{q'-1}^{\mu'},$$

where b, b', b'', μ, μ' are certain integers. By the relation (2.18), we have

$$1 = (\zeta_{q-1}^{\mu})^{\tau-1} = (\zeta_{q'-1}^{b})^{1+\phi+\cdots+\phi^{z-1}} = \zeta_{q-1}^{b},$$

$$1 = (\zeta_{q-1}^{\mu})^{\iota-1} = (\zeta_{q'-1}^{b'})^{1+\phi+\cdots+\phi^{z-1}} = \zeta_{q-1}^{b'},$$

$$1 = (\zeta_{q'-1}^{\mu'})^{\tau-1} = (\zeta_{q'-1}^{-b''})^{1+\iota} = \zeta_{q'-1}^{-2b''},$$

whence $(q-1) \mid b$, $(q-1) \mid b'$, $(q'-1) \mid b''$, $((2, q'-1) = 1)$. Therefore, $v_\iota v_\tau = v_\tau v_\iota$, and there exist integers X, X' satisfying $(q-1)X \equiv b \pmod{q'-1}$, $(q-1)X' \equiv b' \pmod{q'-1}$. Set $w_\tau = \zeta_{q'-1}^{X} v_\tau$, $w_\iota = \zeta_{q'-1}^{X'} v_\iota$. Then

$$v_\phi w_\tau = \zeta_{q'-1}^{Xq-b} v_\tau v_\phi = \zeta_{q'-1}^{X} v_\tau v_\phi = w_\tau v_\phi,$$

$$v_\phi w_\iota = \zeta_{q'-1}^{X'q-b'} v_\iota v_\phi = \zeta_{q'-1}^{X'} v_\iota v_\phi = w_\iota v_\phi,$$

$$w_\tau w_\iota = \zeta_{q'-1}^{X} v_\tau \zeta_{q'-1}^{X'} v_\iota = w_\iota w_\tau.$$

Hence we have

$$B' = \sum_{i=0}^{e/2-1} \sum_{j=0}^{1} \sum_{\nu=0}^{z-1} \Omega \cdot F_1 \cdot F_2 w_\tau^i w_\iota^j v_\phi^\nu$$

$$\simeq (w_\tau^{e/2}, F_1/k, \tau) \otimes_k (w_\iota^2, F_2/k, \iota) \otimes_k (v_\phi^z, \Omega/k, \phi),$$

where $w_\tau^{e/2}$, w_ι^2, v_ϕ^z lie in $\mathcal{W}'(k) = \langle\zeta_{q-1}\rangle$. Thus the same argument as before yields that all the above cyclic algebras are similar to k, and hence $B \sim B' \sim k$. #

We will frequently use the following wellknown fact.

Lemma 5.2. Let m be a positive integer. Write $m = 2^c m'$, $c \geqq 0$, $(2, m') = 1$. Then $5^m \equiv 1 \pmod{2^{c+2}}$, $5^m \not\equiv 1 \pmod{2^{c+3}}$, i.e., $5^m - 1$ is exactly divisible by 2^{c+2}.

Proof. See, for instance, [29,§4, 5].

We will prove that if the inertia group $\mathcal{T}$ of the extension L/k is either of type (ii) or (iii), then $B = (\beta, L/k) \sim k$.

Theorem 5.3 (Yamada [54]). Let $B = (\beta, L/k)$ be a cyclotomic algebra over k defined by (5.1)-(5.5). Let $\mathcal{T}$ denote the inertia group of L/k. If $\mathcal{T} = \langle\theta^{2^\lambda}\rangle$ $(0 \leqq \lambda \leqq n-2)$, or if $\mathcal{T} = \langle\iota\theta^{2^\nu}\rangle$ $(0 \leqq \nu \leqq n-3)$, then $B \sim k$.

Proof. Let the notation be as in (5.1)-(5.22). By Proposition 5.1 we may assume $\beta(\sigma, \tau) \in \langle\zeta_{2^n}\rangle = \mathcal{W}_2(L)$ for any $\sigma, \tau \in \mathcal{G}(L/k)$.

(I) The case $\mathcal{T}' \simeq \mathcal{T} = \langle\theta^{2^\lambda}\rangle$, $(0 \leqq \lambda \leqq n-2)$. Then

$e = e_{L'/k} = e_{L/k} = 2^{n-2-\lambda}$. Set $\tau = \theta^{2^\lambda}$, $(\tau^e = 1)$. By (5.22) we have

$$B \sim B' = \sum_{i=0}^{e-1} \sum_{j=0}^{z-1} \Omega \cdot F v_\tau^i v_\phi^j, \quad (v_1 = 1). \tag{5.25}$$

Let $\beta(\tau, \eta)/\beta(\eta, \tau) = \beta'(\tau, \phi)/\beta'(\phi, \tau) = \zeta_{2^n}^b$, so

$$v_\tau v_\phi = \zeta_{2^n}^b v_\phi v_\tau. \tag{5.26}$$

By (5.20), v_τ^e lies in $\langle \zeta_{2^n} \rangle$. Since $v_\tau v_\tau^e v_\tau^{-1} = v_\tau^e$ and $5^{2^\lambda} - 1$ is exactly divisible by $2^{\lambda+2}$, it follows readily that

$$v_\tau^e = \zeta_{2^{\lambda+2}}^c, \quad c \text{ being some integer.} \tag{5.27}$$

Suppose first that $y = 1$ in (5.14) or (5.15), so $\phi = (\xi')^{f^*} \theta^{x_1}$. By the relation (2.18) we have

$$\zeta_{2^{\lambda+2}}^{cA} = (\zeta_{2^{\lambda+2}}^c)^{\phi-1} = \zeta_{2^n}^{-b(1+\tau+\cdots+\tau^{e-1})} = \zeta_{2^n}^{-bS}, \tag{5.28}$$

$$A = -5^x - 1, \quad S = 1+5^{2^\lambda}+\cdots+5^{2^\lambda(e-1)} = (5^{2^{n-2}}-1)/(5^{2^\lambda}-1). \tag{5.29}$$

The number S (resp. A) is exactly divisible by $2^{n-2-\lambda}$ (resp. 2). By (5.28) we conclude that $2 \mid b$. Let Y be an integer satisfying $AY \equiv b \pmod{2^n}$. (Since $(2, A/2) = 1$ and $2 \mid b$, such an integer Y does exist.) Set $w_\tau = \zeta_{2^n}^Y v_\tau$. Then

$$v_\phi w_\tau = \zeta_{2^n}^{-5^x Y-b} v_\tau v_\phi = \zeta_{2^n}^{Y} v_\tau v_\phi = w_\tau v_\phi .$$

It follows from (5.25) that

$$B \sim B' = \sum_{i=0}^{e-1} \sum_{j=0}^{z-1} \Omega\cdot F w_\tau^i v_\phi^j$$

$$\approx (w_\tau^e, F/k, \tau) \otimes_k (v_\phi^z, \Omega/k, \phi)$$

$$\sim (w_\tau^e, F/k, \tau),$$

where $w_\tau^e = \zeta_{2^n}^{Y(1+\tau+\cdots+\tau^{e-1})} v_\tau^e \in W_2(k)$, $v_\phi^z \in W_2(k)$, and Ω/k is unramified. Note that $\zeta_4^\phi = \zeta_4^{-1}$, so $\zeta_4 \notin k$, and $W_2(k) =$ $\langle -1\rangle$. Hence $B' \sim (\pm 1, F/k, \tau)$. Because $e_{k/Q_2} = 2^{n-1}/e = 2^{1+\lambda}$ $\equiv 0 \pmod 2$, it follows that $2 \mid [k : Q_2]$, $N_{k/Q_2}(-1) = 1$, and the order of the norm residue symbol $(-1, F/k) = (N_{k/Q_2}(-1), F/Q_2) = (1, F/Q_2)$ is equal to 1. Thus we have $B \sim B' \sim k$, as required.

Suppose next that $y = 0$ in (5.15), so $\phi = (\xi')^{f^*}\theta^x$. We may assume $1 \leqq x \leqq 2^{n-2}$. Write $x = 2^\kappa t$, $0 \leqq \kappa \leqq n-2$, $(2, t) = 1$. By the relation (2.18) we have

$$\zeta_{2^{\lambda+2}}^{cI} = (\zeta_{2^{\lambda+2}}^{c})^{\phi-1} = \zeta_{2^n}^{-b(1+\tau+\cdots+\tau^{e-1})} = \zeta_{2^n}^{-bS}, \tag{5.30}$$

where $I = 5^x - 1$, and S is defined by (5.29). Since I (resp. S) is exactly divisible by $2^{\kappa+2}$ (resp. $2^{n-2-\lambda}$), it follows easily from (5.30) that

(i) if $\kappa < \lambda$, then $2^{\kappa+2} \mid b$; (5.31)

(ii) if $\kappa \geq \lambda$, then $2^{\lambda+2} \mid b$. (5.32)

If $\kappa < \lambda$, we let X be an integer satisfying $IX \equiv b \pmod{2^n}$. (By (5.31), such an integer X does exist.) Set $w_\tau = \zeta_{2^n}^X v_\tau$. Then it is easily verified that $v_\phi w_\tau = w_\tau v_\phi$, and so by (5.25) we have

$$B \sim B' = \sum_{i=0}^{e-1} \sum_{j=0}^{z-1} \Omega \cdot F w_\tau^i v_\phi^j$$

$$\approx (w_\tau^e, F/k, \tau) \otimes_k (v_\phi^z, \Omega/k, \phi)$$

$$\sim (w_\tau^e, F/k, \tau), \qquad w_\tau^e \in \mathcal{W}_2(k). \tag{5.33}$$

Since $G(L'/k) = \langle\theta^{2^\lambda}\rangle \times \langle(\xi')^{f^*}\theta^x\rangle$, $x = 2^\kappa t$, $\kappa < \lambda$, it follows that $\mathcal{W}_2(k) = \langle\zeta_{2^{\kappa+2}}\rangle$. The inertia group T'' of the extension k/Q_2 is canonically isomorphic to the factor group $((\langle\theta\rangle \times \langle\iota\rangle)\cdot G(L'/k))/G(L'/k)$, whose complete set of representatives are $\theta^i\iota^j$ $(0 \leq i \leq 2^\lambda-1;\ j = 0,1)$. Denote by k_0 the subfield of k over Q_2 corresponding to T'' in the sense of Galois theory, i.e., k_0 is the inerta field in k/Q_2. Then

$$N_{k/k_0}(\zeta_{2^{\kappa+2}}) = \prod_{i=0}^{2^\lambda-1} \prod_{j=0}^{1} \zeta_{2^{\kappa+2}}^{\theta^i\iota^j} = 1, \tag{5.34}$$

because for each i $(i = 0,1,\cdots,2^\lambda-1)$, we have

$$\zeta_{2^{\kappa+2}}^{\theta^i+\iota\theta^i} = \zeta_{2^{\kappa+2}}^{5^i-5^i} = 1.$$

Therefore, $N_{k/Q_2}(\zeta_{2^{\kappa+2}}) = 1$ and hence the order of the norm residue symbol $(w_\tau^e,\ F/k) = (N_{k/Q_2}(w_\tau^e),\ F/Q_2) = (1,\ F/Q_2)$ is equal to 1, $(w_\tau^e \in \langle\zeta_{2^{\kappa+2}}\rangle)$. Thus, by (5.33) we have $B \sim B' \sim k$, as requested.

If $\kappa \geqq \lambda$, we let X' be an integer satisfying $(5^{2^\lambda} - 1)X' \equiv -b \pmod{2^n}$. (By (5.32), such an integer exists.) Set $w_\phi = \zeta_{2^n}^{X'} v_\phi$. Then it follows from (5.26) that

$$v_\tau w_\phi = \zeta_{2^n}^{X'5^{2^\lambda}+b} v_\phi v_\tau = \zeta_{2^n}^{X'} v_\phi v_\tau = w_\phi v_\tau,$$

whence by (5.25) we have

$$\begin{aligned} B \sim B' &= \sum_{i=0}^{e-1} \sum_{j=0}^{z-1} \Omega\cdot F v_\tau^i w_\phi^j \\ &\cong (v_\tau^e,\ F/k,\ \tau) \otimes_k (w_\phi^z,\ \Omega/k,\ \phi) \\ &\sim (v_\tau^e,\ F/k,\ \tau), \end{aligned}$$

$$v_\tau^e \in \omega_2(k), \quad w_\phi^z = \zeta_{2^n}^{X'(1+\phi+\cdots+\phi^{z-1})} v_\phi^z \in \omega_2(k).$$

Since $\kappa \geqq \lambda$, we have $\omega_2(k) = \langle\zeta_{2^{\lambda+2}}\rangle$. The notation being as in the case $\kappa < \lambda$, we have

$$N_{k/k_0}(\zeta_{2^{\lambda+2}}) = \prod_{i=0}^{2^\lambda-1} \prod_{j=0}^{1} \zeta_{2^{\lambda+2}}^{\theta^i \iota^j} = 1.$$

Hence $N_{k/Q_2}(\zeta_{2^{\lambda+2}}) = 1$, and so $N_{k/Q_2}(v_\tau^e) = 1$. Thus, $B \sim B' \sim k$, as required.

(II) The case $T' \simeq T = \langle \iota\theta^{2^\nu} \rangle$, $(0 \leq \nu \leq n-3)$. Set $\tau = \iota\theta^{2^\nu}$. Then $e = e_{L'/k} = 2^{n-2-\nu}$, $\tau^e = 1$. Since $\zeta_4^\tau = \zeta_4^{-1}$, and $v_\tau v_\tau^e v_\tau^{-1} = v_\tau^e$, it follows from (5.22), (5.19), (5.20) that

$$B \sim B' = (\beta', L'/k) = \sum_{i=0}^{e-1} \sum_{\nu=0}^{z-1} \Omega \cdot F v_\tau^i v_\phi^\nu, \tag{5.35}$$

$$v_\tau v_\phi = \zeta_{2^n}^b v_\phi v_\tau, \qquad v_\tau^e = (-1)^c,$$

where b, c are certain integers. By the relation (2.18) we conclude that

$$1 = ((-1)^c)^{\phi-1} = (\zeta_{2^n}^{-b})^{1+\tau+\cdots+\tau^{c-1}} = \zeta_{2^n}^{-bJ},$$

$$J = 1+(-5^{2^\nu})+ \cdots +(-5^{2^\nu})^{e-1} = (1-5^{2^{n-2}})/(1+5^{2^\nu}).$$

The number J is exactly divisible by 2^{n-1}, so $2 \mid b$, Let X be an integer satisfying $(1 + 5^{2^\nu})X \equiv b \pmod{2^n}$. Put $w_\phi = \zeta_{2^n}^X v_\phi$. Then

$$v_\tau w_\phi = \zeta_{2^n}^{X(-5^{2^\nu})+b} v_\phi v_\tau = \zeta_{2^n}^X v_\phi v_\tau = w_\phi v_\tau.$$

It follows from (5.35) that

$$B' = \sum_{i=0}^{e-1} \sum_{\nu=0}^{z-1} \Omega \cdot F v_\tau^i w_\phi^\nu$$

$$\simeq (v_\tau^e,\ F/k,\ \tau) \otimes_k (w_\phi^z,\ \Omega/k,\ \phi) \sim ((-1)^c,\ F/k,\ \tau),$$

($w_\phi^z \in \mathcal{W}_2(k)$, Ω/k is unramified). Note that $N_{k/Q_2}(-1) = 1$, because $e_{k/Q_2} = 2^{1+\nu}$. Thus, $((-1)^c, F/k) = (N_{k/Q_2}((-1)^c), F/Q_2) = 1 \in G(F/Q_2)$, and $B \sim B' \sim k$. This completes the proof of Theorem 5.3.

Corollary 5.4. If $\zeta_4 \in k$, Then $S(k) = 1$.

Proof. Let $B = (\beta, L/k)$ be a cyclotomic algebra over k, and T the inertia group of L/k. Let the notation be as before. We notice that

$$\zeta_4^\iota = \zeta_4^{\iota\theta^{2^\nu}} = \zeta_4^{-1} \neq \zeta_4.$$

Hence $T = \langle\theta^{2^\lambda}\rangle$ for some λ. Then Theorem 5.3 implies that $B = (\beta, L/k) \sim k$. Since B is an arbitrary cyclotomic algebra over k, it follows that $S(k) = 1$. #

Corollary 5.5. Let k be a cyclotomic extension of Q_2. Then the order of $S(k)$ is 1 or 2. In other words, if $[A] \in S(k)$, then

$$\mathrm{inv}_k(A) \equiv 0 \text{ or } \frac{1}{2} \pmod{Z}.$$

Proof. By Corollary 5.4, $S(k(\zeta_4)) = 1$. If $[A] \in S(k)$, then $[A \otimes k(\zeta_4)] \in S(k(\zeta_4)) = 1$. Hence $k(\zeta_4)$ is a splitting field of A. Since $[k(\zeta_4) : k] = 1$ or 2, we conclude that $\mathrm{inv}_k(A) \equiv 0$ or $\frac{1}{2}$ (mod Z). #

Next we will give a formula of invariant of a 2-adic cyclotomic algebra $(\beta, L/k)$, when the inertia group T of L/k is "non-cyclic".

Theorem 5.6 (Yamada [54]). Let $B = (\beta, L/k) = \sum_\sigma Lu_\sigma$ be a cyclotomic algebra over k given by (5.1)-(5.5), and such that $\beta(\sigma, \tau) \in \langle\zeta_{2^n}\rangle$ for all $\sigma, \tau \in G(L/k)$. Let the notation be as in (5.7)-(5.22). Let T denote the inertia group of L/k, and suppose $T = \langle\theta^{2^\lambda}\rangle \times \langle\iota\rangle$, $(0 \leqq \lambda \leqq n - 2)$. Let η be a Frobenius automorphism of L/k. It may be assumed that $\eta = \zeta^{f^*}\theta^x$, $1 \leqq x \leqq 2^{n-2}$, $x = 2^\kappa t$, $(2, t) = 1$, $0 \leqq \kappa \leqq n - 2$, $(f^* = f_{k/Q_2})$. Put $\tau = \theta^{2^\lambda}$, and let

$$\beta(\tau, \eta)/\beta(\eta, \tau) = \zeta_{2^n}^{a}, \quad \beta(\iota, \eta)/\beta(\eta, \iota) = \zeta_{2^n}^{a'}, \qquad (5.36)$$

$$\beta(\tau, \iota)/\beta(\iota, \tau) = \zeta_{2^n}^{a''}. \qquad (5.37)$$

Put

$$b = \begin{cases} (2a + (5^{2^\lambda}-1)a' + (5^x-1)a'')/2^n & \text{for } 0 \leqq \lambda < n-2, \\ -a' & \text{for } \lambda = n - 2 \quad (T = \langle\iota\rangle). \end{cases} \qquad (5.38)$$

Then b is an integer. If either $T = \langle \iota \rangle$ $(\lambda = n - 2)$ or $2 \mid a'$, then the invariant of $B = (\beta, L/k)$ is equal to:

$$\mathrm{inv}_k(B) \equiv \frac{b}{2} \pmod{Z} \tag{5.39}$$

except the case $2 \nmid [k : Q_2]$, $\beta(\iota, \iota) = -1$, for which

$$\mathrm{inv}_k(B) \equiv \frac{b+1}{2} \pmod{Z}. \tag{5.40}$$

If either $\lambda \leq \kappa$ $(\lambda \neq n - 2)$, or the 2-part of $f \leq 2^{n-2-\kappa}$, then $2 \mid a'$. (f is the residue class degree of L/k.)

Remark 5.7. Write $\eta = \xi^{f^*}\theta^x\iota^y$ $(1 \leq x \leq 2^{n-2};\ y = 0, 1)$ as in (5.14). Note that for any element τ of T, $\eta\tau$ is also a Frobenius automorphism of L/k. So we may assume $\eta = \xi^{f^*}\theta^x$, because $\iota \in T$. If $\lambda \leq \kappa$, then $\xi^{f^*} = \eta(\theta^{2^\lambda})^{-2^{\kappa-\lambda}t} \in G(L/k)$, and hence we may assume $\eta = \xi^{f^*}$. In particular, if k/Q_2 is unramified, i.e., if $T = \langle \theta \rangle \times \langle \iota \rangle$ $(\lambda = 0)$, this is always the case. Note that if $n = 3$ and T is "non-cyclic", then either $T = \langle \iota \rangle$ or $T = \langle \theta \rangle \times \langle \iota \rangle$.

Proof of Theorem 5.6. Notation being the same as in (5.1)-(5.22) we have

$$e = e_{L/k} = e_{L'/k}, \quad G(L'/k) = \langle \tau \rangle \times \langle \iota \rangle \times \langle \phi \rangle,$$

$$\phi = (\xi')^{f^*}\theta^x, \quad \zeta_{2^n}^{\phi} = \zeta_{2^n}^{5^x}.$$

Suppose that $\tau \neq 1$, i.e., $\lambda < n - 2$. Then $e = 2^{n-\lambda-1}$. For simplicity, put $e' = e/2 = 2^{n-\lambda-2}$. By (5.19)-(5.22) we have

$$B = (\beta, L/k) \sim B' = (\beta', L'/k)$$

$$= \sum_{i=0}^{e'-1} \sum_{j=0}^{1} \sum_{\nu=0}^{z-1} L' v_\tau^i v_\iota^j v_\phi^\nu \qquad (5.41)$$

$$v_\tau v_\phi = \zeta_{2^n}^{a} v_\phi v_\tau, \quad v_\iota v_\phi = \zeta_{2^n}^{a'} v_\phi v_\iota, \quad v_\tau v_\iota = \zeta_{2^n}^{a''} v_\iota v_\tau, \qquad (5.42)$$

$$v_\tau^{e'} = \zeta_{2^{\lambda+2}}^{c}, \quad v_\iota^2 = \beta(\iota, \iota) = \pm 1, \quad v_\phi^z \in \langle \zeta_{2^n} \rangle, \qquad (5.43)$$

where a, a', a'' are given by (5.36), (5.37), and c is a certain integer (determined by β). From the relation (2.18) we conclude that

$$\zeta_{2^{\lambda+2}}^{-2c} = (\zeta_{2^{\lambda+2}}^{c})^{\iota-1} = (\zeta_{2^n}^{-a''})^{1+\tau+\cdots+\tau^{e'-1}} = \zeta_{2^n}^{-a''S}, \qquad (5.44)$$

$$\zeta_{2^{\lambda+2}}^{5^x-1} = (\zeta_{2^{\lambda+2}}^{c})^{\phi-1} = (\zeta_{2^n}^{-a})^{1+\tau+\cdots+\tau^{e'-1}} = \zeta_{2^n}^{-aS}, \qquad (5.45)$$

$$S = 1 + 5^{2^\lambda} + \cdots + (5^{2^\lambda})^{e'-1} = (5^{2^{n-2}} - 1)/(5^{2^\lambda} - 1). \quad (5.46)$$

It follows from Lemma 5.2 that

$$2^{n-\lambda-2} \,\|\, S, \qquad 2^{\kappa+2} \,\|\, 5^x - 1, \qquad (5.47)$$

$(x = 2^\kappa t,\ (2, t) = 1)$. Hence by (5.44), (5.45), we conclude

$$2 \mid a'', \quad \text{and if} \quad \lambda \leqq \kappa, \quad \text{then} \quad 2^{\lambda+2} \mid a. \tag{5.48}$$

By the relation (2.19) and (5.42), we have

$$(\zeta_{2^n}^{a})^{\iota-1}(\zeta_{2^n}^{-a'})^{\tau-1}(\zeta_{2^n}^{-a''})^{\phi-1} = \zeta_{2^n}^{-2a-(5^{2^\lambda}-1)a'-(5^x-1)a''} = 1,$$

whence

$$2a + (5^{2^\lambda} - 1)a' + (5^x - 1)a'' \equiv 0 \pmod{2^n}. \tag{5.49}$$

Thus the number b in (5.38) is an integer.

From now on we assume $2 \mid a'$. Put

$$w_\tau = \zeta_{2^n}^{-a''/2} v_\tau, \qquad w_\phi = \zeta_{2^n}^{d} v_\phi \tag{5.50}$$

$$d = a'/2 + 2^{n-\lambda-3} b. \tag{5.51}$$

Then it follows from (5.42) that

$$w_\tau v_\iota = \zeta_{2^n}^{-a''/2+a''} v_\iota v_\tau = \zeta_{2^n}^{a''/2} v_\iota v_\tau = v_\iota(\zeta_{2^n}^{-a''/2} v_\tau) = v_\iota w_\tau, \tag{5.52}$$

and that

$$w_\tau w_\phi = \zeta_{2^n}^{-a''/2} v_\tau \zeta_{2^n}^{d} v_\phi = \zeta_{2^n}^{-a''/2+d5^{2^\lambda}+a} v_\phi v_\tau$$

$$= \zeta_{2^n}^{-a''/2+d5^{2^\lambda}+a-d+5^x a''/2}(\zeta_{2^n}^{d} v_\phi)(\zeta_{2^n}^{-a''/2} v_\tau)$$

$$= \zeta_{2^n}^{a+(5^{2^\lambda}-1)d+(5^x-1)a''/2} w_\phi w_\tau = w_\phi w_\tau, \tag{5.53}$$

because by (5.51), (5.38), we have

$$a + (5^{2^\lambda}-1)d + (5^x-1)a''/2$$

$$= \{2a + 2(5^{2^\lambda}-1)(a'/2+b2^{n-\lambda-3}) + (5^x-1)a''\}/2$$

$$= \{b2^n + (5^{2^\lambda}-1)b2^{n-\lambda-2}\}/2 = \{b2^n + b2^n(5^{2^\lambda}-1)/2^{\lambda+2}\}/2$$

$$= b2^{n-1}(1 + (5^{2^\lambda}-1)/2^{\lambda+2}) \equiv 0 \pmod{2^n}.$$

In the above, we recall that $(5^{2^\lambda} - 1)/2^{\lambda+2}$ is an odd integer. Thus w_τ commutes with v_ι and w_ϕ. Let K (resp. F') be the subfield of L' over k corresponding to $\langle\tau\rangle$ (resp. $\langle\iota\rangle \times \langle\phi\rangle$) in the sense of Galois theory. Then $L' = K\cdot F'$, $K \cap F' = k$, $G(K/k) \simeq \langle\iota\rangle \times \langle\phi\rangle$, $G(F'/k) \simeq \langle\tau\rangle$. By (5.41), (5.53), (5.50) we have

$$B \sim B' = \sum_{i=0}^{e'-1} \sum_{j=0}^{1} \sum_{\nu=0}^{z-1} K\cdot F' w_\tau^i v_\iota^j w_\phi^\nu$$

$$= [\sum_{i=0}^{e'-1} F' w_\tau^i]\cdot[\sum_{j=0}^{1} \sum_{\nu=0}^{z-1} K v_\iota^j w_\phi^\nu]$$

$$\simeq (w_\tau^{e'}, F'/k, \tau) \otimes_k (\beta' | (\langle\iota\rangle\times\langle\phi\rangle), K/k), \tag{5.54}$$

$$w_\tau^{e'} = \zeta_{2^n}^{(-a''/2)(1+\tau+\cdots+\tau^{e'-1})} v_\tau^{e'} \in W_2(k) = \{\pm 1\}.$$

(Because $G(L'/k) \ni \iota$, $\zeta_4^{\iota} = \zeta_4^{-1}$, it follows that $w_2(k) = \{\pm 1\}$.) Obviously, $(1, F'/k, \tau) \sim k$. If $2 \mid [k : Q_2]$, then the norm residue symbol $(-1, F'/k) = (N_{k/Q_2}(-1), F'/Q_2) = 1 \in G(F'/Q_2)$, and so $(-1, F'/k, \tau) \sim k$. If $2 \nmid [k : Q_2]$, then k/Q_2 is unramified and hence $G(L'/k) = \langle\theta\rangle \times \langle\iota\rangle \times \langle\phi\rangle$, $(\lambda = 0, \tau = \theta)$. In this case $F' = k \cdot k_0$, $k_0 = Q_2(\zeta_{2^n} + \zeta_{2^n}^{-1})$, $k \cap k_0 = Q_2$, $\langle\theta\rangle \simeq G(F'/k) \simeq G(k_0/Q_2)$. Hence we have

$$(-1, F'/k, \theta) \simeq (-1, k_0/Q_2, \theta) \otimes_{Q_2} k.$$

But it is wellknown that $(-1, k_0/Q_2, \theta) \sim Q_2$, whence $(-1, F'/k, \theta) \sim k$. (For instance, we can explain this fact as follows: Set $\rho = (-1, Q_2(\zeta_{2^n})/Q_2) \in G(Q_2(\zeta_{2^n})/Q_2)$. Then k_0 is exactly the fixed field of $\langle\rho\rangle$, whence $(-1, k_0/Q_2) = \rho|k_0 = 1$.) Thus, in any case we have $(\pm 1, F'/k, \tau) \sim k$, and consequently, by (5.54), (5.50), (5.42),

$$B \sim B' \sim (\beta'|(\langle\iota\rangle\times\langle\phi\rangle), K/k) = \sum_{j=0}^{1} \sum_{\nu=0}^{z-1} K v_\iota^j w_\phi^\nu \tag{5.55}$$

$$v_\iota w_\phi = \zeta_{2^n}^{-d+a'} v_\phi v_\iota = \zeta_{2^n}^{-2d+a'} w_\phi v_\iota . \tag{5.56}$$

From (5.51) we have

$$-2d + a' = -a' - b2^{n-\lambda-2} + a' = -b2^{n-\lambda-2},$$

and hence

$$\zeta_{2^n}^{-2d+a'} = \zeta_{2^n}^{-2^{n-\lambda-2}b} = \zeta_{2^{\lambda+2}}^{-b}, \quad (\zeta_{2^n}^{2^{n-\lambda-2}} = \zeta_{2^{\lambda+2}}),$$

$$v_\iota w_\phi = \zeta_{2^{\lambda+2}}^{-b} w_\phi v_\iota. \tag{5.57}$$

Thus under the assumption $\lambda < n - 2$ $(\tau \neq 1)$, $2 \mid a'$, we have obtained (5.55), (5.57). However, (5.55) and (5.57) also hold for the case $\lambda = n - 2$ $(T = \langle\iota\rangle)$, if we put

$$K = L', \quad d = 0, \quad w_\phi = v_\phi. \tag{5.58}$$

(See (5.38).) Hence we need only compute the invariant of the cyclotomic algebra in (5.55) with the relation (5.57).

Set

$$y_\phi = (1 + \zeta_{2^{\lambda+2}})^{-b} w_\phi.$$

Then it follows from (5.57) that

$$v_\iota y_\phi = (1 + \zeta_{2^{\lambda+2}}^{-1})^{-b} \zeta_{2^{\lambda+2}}^{-b} w_\phi v_\iota = y_\phi v_\iota. \tag{5.59}$$

Let E (resp. F) denote the fixed field of $\langle\iota\rangle$ (resp. $\langle\phi\rangle$) in K/k. Then $K = E \cdot F$, $G(E/k) \simeq \langle\phi\rangle$, $G(F/k) \simeq \langle\iota\rangle$. From (5.55), (5.59), (5.50) we have

$$B \sim B' \sim \sum_{j=0}^{1} \sum_{\nu=0}^{z-1} E \cdot F v_\iota^j y_\phi^\nu$$

$$\simeq (v_\iota^2, F/k, \iota) \otimes_k (y_\phi^z, E/k, \phi), \tag{5.60}$$

$$y_\phi^z = \{(1+\zeta_{2^{\lambda+2}})^{-b}\zeta_{2^n}^d v_\phi\}^z$$

$$= \{\prod_{i=0}^{z-1} (1+\zeta_{2^{\lambda+2}}^{\phi^i})^{-b}(\zeta_{2^n}^d)^{\phi^i}\}\cdot v_\phi^z, \quad (v_\phi^z \in \mathcal{W}_2(L')). \tag{5.61}$$

Let V_k (resp. $V_{L'}$) denote the normalized discrete valuation of k (resp. L'), i.e., $V_k(\pi) = V_{L'}(\pi') = 1$, where π (resp. π') is a prime element of k (resp. L'). Because

$$1 + \zeta_{2^{\lambda+2}}^{\phi^i} \qquad (i = 0, 1, \cdots, z-1)$$

is a prime element of $Q_2(\zeta_{2^{\lambda+2}})$, and the ramification index of the extension $L'/Q_2(\zeta_{2^{\lambda+2}})$ (resp. L'/k) equals $2^{n-\lambda-2}$ (resp. $2^{n-\lambda-1}$), we conclude from (5.61) that

$$2^{n-\lambda-1}\cdot V_k(y_\phi^z) = V_{L'}(y_\phi^z) = V_{L'}(\prod_{i=0}^{z-1} (1+\zeta_{2^{\lambda+2}}^{\phi^i})^{-b}) = 2^{n-\lambda-2}\cdot z(-b),$$

and so

$$V_k(y_\phi^z) = \frac{z(-b)}{2}. \tag{5.62}$$

Because E/k is an unramified extension of degree z, it follows from the definition of Hasse invariant and from (5.62) that

$$\mathrm{inv}_k((y_\phi^z, E/k, \phi)) = V_k(y_\phi^z)/z = \frac{-b}{2} \equiv \frac{b}{2} \pmod{Z}. \tag{5.63}$$

Next we consider the cyclic algebra $(v_\iota^2, E/k, \iota)$ in (5.60),

($v_\iota^2 = \beta(\iota, \iota) = \pm 1$), whose index is equal to the order of the norm residue symbol $(\pm 1, F/k) = (N_{k/Q_2}(\pm 1), F/Q_2)$. If $2 | [k : Q_2]$, then $N_{k/Q_2}(\pm 1) = 1$, and hence $(v_\iota^2, F/k, \iota) \sim k$. If $2 \nmid [k : Q_2]$, then it is easy to see that k/Q_2 is unramified, $F = k(\zeta_4)$, and $N_{k/Q_2}(-1) = -1$. But there is no element $\delta \in F$ such that $N_{F/Q_2}(\delta) = -1$, because there is no element $\delta' \in Q_2(\zeta_4) \subset k(\zeta_4) = F$ such that $N_{Q_2(\zeta_4)/Q_2}(\delta') = -1$. Hence the order of the norm residue symbol $(-1, F/Q_2)$ equals 2. Thus

$$\mathrm{inv}_k((v_\iota^2, F/k, \iota)) = \begin{cases} \frac{1}{2}, & \text{if } 2 \nmid [k : Q_2],\ \beta(\iota, \iota) = -1, \\ 0, & \text{otherwise.} \end{cases} \tag{5.64}$$

The equations (5.39), (5.40) readily follow from (5.60), (5.63), (5.64).

Finally we will prove the last statement of Theorem 5.4. Suppose $\lambda \leqq \kappa$, $(\lambda \neq n - 2)$. Then $2^{\lambda+2} \parallel 5^{2^\lambda} - 1$, $2^{\lambda+2} \mid 5^x - 1$. Hence we conclude from (5.48), (5.49) that $2|a'$.

Denote by 2^μ the 2-part of f, $(f = f_{L/k})$, and suppose $2^\mu \leqq 2^{n-2-\kappa}$. Since

$$\eta = \xi^{f^*}\theta^x, \quad x = 2^\kappa t, \quad \eta^f = \xi^{f^* f}\theta^{xf} = \theta^{xf},$$

we see that the 2-part of the order of η is equal to $2^{n-2-\kappa}$. Put $z = 2^{n-2-\kappa}\cdot f/2^\mu$. Let Ω denote the unramified extension of k of degree z. Then a similar argument to the one given in (5.11)-(5.16) yields that the composite field $L' = L\cdot\Omega$ has a Frobenius automorphism of order $z = f_{L\cdot\Omega/k}$, the residue class degree of L'/k. Hence we can use the same notation as in (5.13)-(5.22), (5.41)-(5.43). In particular,

$$L' = L\cdot\Omega, \quad \mathcal{G}(L'/k) = \langle\theta^{2^\lambda}\rangle \times \langle\iota\rangle \times \langle\phi\rangle, \quad \phi = (\xi')^{f^*}\theta^x,$$

$$\phi|L = \eta, \quad \phi^z = 1, \quad z = 2^{n-2-\kappa}\cdot f/2^\mu.$$

Since

$$\zeta_{2^n}^{\phi} = \zeta_{2^n}^{5^x}, \quad x = 2^\kappa t, \quad (2, t) = 1, \quad (v_\phi^z)^\phi = v_\phi^z,$$

it follows from Lemma 5.2 that $v_\phi^z = \zeta_{2^{\kappa+2}}^{c'}$ for some integer c'. By the relation (2.18) and (5.42) we have

$$\zeta_{2^{\kappa+2}}^{-2c'} = (\zeta_{2^{\kappa+2}}^{c'})^{\iota-1} = (\zeta_{2^n}^{a'})^{1+\phi+\cdots+\phi^{z-1}} = \zeta_{2^n}^{a'S'}, \tag{5.65}$$

$$S' = 1 + 5^x + \cdots + 5^{x(z-1)} = (5^{2^\kappa t 2^{n-2-\kappa}\cdot f/2^\mu} - 1)/(5^{2^\kappa t} - 1)$$
$$= (5^{2^{n-2}t\cdot f/2^\mu} - 1)/(5^{2^\kappa t} - 1). \tag{5.66}$$

Since $t\cdot f/2^\mu$ is an odd integer, it follows from (5.66) that $2^{n-\kappa-2} \| S'$, and consequently, from (5.65) we conclude that $2|a'$. This completes the proof of Theorem 5.6.

By Proposition 5.1, Theorem 5.3, and Theorem 5.6 we can determine the Schur subgroup of a 2-adic field. Let h be the least non-negative integer such that k is contained in a cyclotomic field $Q_2(\zeta_{2^h}, \zeta_{2^b-1})$ for some integer b. We will call h the <u>height</u> of k. It is obvious that either $h = 0$ or $h \geq 2$, and that $h = 0$ if and only if k/Q_2 is unramified.

<u>Lemma 5.8</u>. Let h be the height of k. Set $M = k(\zeta_{2^h})$. Then M is a cyclotomic field (over Q_2) and contained in every cyclotomic field which contains k. That is, M is the minimal cyclotomic field containing k. If the residue class degree of M/Q_2 is f then $M = Q_2(\zeta_{2^h}, \zeta_{2^f-1})$.

<u>Proof</u>. If $h = 0$, the assertions are clear. Suppose that $h \geq 2$. Let the residue class degree of k/Q_2 be f^* and set $K = Q_2(\zeta_{2^h}, \zeta_{2^{f^*}-1})$. Then $\zeta_{2^{f^*}-1} \in k$, and so $k(\zeta_{2^h}) \supset K$. Let s be an integer such that $L = Q_2(\zeta_{2^h}, \zeta_{2^s-1}) \supset k$. Then $L \supset k(\zeta_{2^h}) \supset K$. Hence f^* divides s and every subfield of L over K is of the form $Q_2(\zeta_{2^h}, \zeta_{2^t-1})$, $f^*|t$, $t|s$. In particular, so is $k(\zeta_{2^h})$. If f is the residue class degree of $k(\zeta_{2^h})/Q_2$, we conclude that $k(\zeta_{2^h}) = Q_2(\zeta_{2^h}, \zeta_{2^f-1})$. If I is a cyclotomic field containing k, then it follows from the definition of the height h of k that $\zeta_{2^h} \in I$, and so $I \supset k(\zeta_{2^h})$. #

Lemma 5.9. Let h be the height of k. Suppose that $h \geqq 2$. Let $L = Q_2(\zeta_{2^c}, \zeta_{2^s-1})$ $(c \geqq h)$ be a cyclotomic field containing k and let E be the maximal unramified extension of k in L. Then $E(\zeta_4) = Q_2(\zeta_{2^h}, \zeta_{2^s-1})$. In particular, if $c = h$ then $E(\zeta_4) = L$. If $k(\zeta_4)/k$ is unramified, then $E = E(\zeta_4)$. If $k(\zeta_4)/k$ is ramified, then $E(\zeta_4)/E$ is also ramified and $E \cap k(\zeta_4) = k$.

Proof. Since the residue class degrees of L and E over Q_2 are the same, ζ_{2^s-1} belongs to E. Recall that the extension $Q_2(\zeta_{2^c})/Q_2(\zeta_4)$ is cyclic and that if I is a subfield of $Q_2(\zeta_{2^c})$ over $Q_2(\zeta_4)$ with $[Q_2(\zeta_{2^c}) : I] = 2^r$ then $I = Q_2(\zeta_{2^{c-r}})$. Since $G(L/Q_2(\zeta_{2^s-1}, \zeta_4))$ is canonically isomorphic to $G(Q_2(\zeta_{2^c})/Q_2(\zeta_4))$ and $E(\zeta_4) \supset Q_2(\zeta_{2^s-1}, \zeta_4)$, we have $E(\zeta_4) = Q_2(\zeta_{2^s-1}, \zeta_{2^{c-t}})$, where $2^t = [L : E(\zeta_4)]$ $(t \geqq 0)$. Hence $E(\zeta_4)$ is a cyclotomic field containing k. It follows from Lemma 5.9 that $E(\zeta_4) \supset M = k(\zeta_{2^h}) = Q_2(\zeta_{2^h}, \zeta_{2^f-1}) \supset k$, $f = f_{M/Q_2}$. Clearly, the ramification index of the extension $E(\zeta_4)/M$ is 2^{c-t-h}. We will prove that $E(\zeta_4)/M$ is unramified, so that $c - t = h$. Suppose first that $k(\zeta_4)/k$ is unramified. Then $\zeta_4 \in E$, because E is the maximal unramified extension of k in L. This implies that $E = E(\zeta_4)$ and $E(\zeta_4)/M$ is unramified.

Suppose next that $k(\zeta_4)/k$ is ramified. Then $\zeta_4 \notin E$ and $E \cap k(\zeta_4) = k$. It is evident that $E(\zeta_4)/E$ is a ramified exten- of degree 2. Since the ramification index of $E(\zeta_4)/k$ is equal to that of $k(\zeta_4)/k$, $E(\zeta_4)/k(\zeta_4)$ is unramified, a fortiori $E(\zeta_4)/M$ is unramified. #

<u>Lemma 5.10</u>. Keeping the notation of Lemma 5.8, suppose that $h \neq 0$ and $k(\zeta_4)/k$ is ramified. Then $h \geqq 3$. Let E be the maximal unramified extension of k in $M = k(\zeta_{2^h})$. Then $M = E(\zeta_4)$ and M/E is ramified. Let

$$\langle\omega\rangle = G(M/E), \quad (\omega^2 = 1), \quad \zeta_{2^h}^{\omega} = \zeta_{2^h}^{z} \tag{5.67}$$

for an integer z. Then either $z \equiv -1 \pmod{2^h}$ or $z \equiv -1 + 2^{h-1} \pmod{2^h}$.

<u>Proof</u>. Assume $h = 2$. Then it follows from Lemma 5.8 that $M = k(\zeta_4) = Q_2(\zeta_4, \zeta_{2^f-1})$, f being the residue class degree of $k(\zeta_4)/Q_2$. Hence, if $k(\zeta_4)/k$ would be ramified, then $k = Q_2(\zeta_{2^f-1})$ and so the height h of k would be equal to 0. This is a contradiction. Thus if $k(\zeta_4)/k$ is ramified, then $h \neq 2$. The second assertion is clear by Lemmas 5.8, 5.9. Recall that $\pm 1 \bmod 2^h$ and $\pm 1 + 2^{h-1} \bmod 2^h$ are just the elements of $Z \bmod^{\times} 2^h$ $(h \geqq 3)$ whose (multiplicative) orders divide 2. As $\omega^2 = 1$, we have $z^2 \equiv 1 \pmod{2^h}$, so that $z \equiv \pm 1$ or $\pm 1 + 2^{h-1} \pmod{2^h}$. Because $M = E(\zeta_4) = E(\zeta_{2^h})$ and $\zeta_4 \notin E$, it follows

that $\zeta_4 \neq \zeta_4^{\omega} = \zeta_4^{z}$, and so $z \not\equiv 1, 1 + 2^{h-1} \pmod{2^h}$. Thus we have either $z \equiv -1$ or $-1 + 2^{h-1} \pmod{2^h}$. #

The Schur subgroup of a 2-adic field is completely determined by the following.

Theorem 5.11 (Yamada [51]). Let k be a cyclotomic extension of Q_2 and let h be the height of k.

(I) Suppose that $k(\zeta_4)/k$ is ramified. Then only the following three cases happen:

(i) $h = 0$,

(ii) $h \geq 3$ and $z \equiv -1 \pmod{2^h}$,

(iii) $h \geq 3$ and $z \equiv -1 + 2^{h-1} \pmod{2^h}$, where z is defined by (5.67).

For the cases (i) and (ii), $S(k)$ is the subgroup of order 2 of $Br(k)$. For the case (iii), $S(k) = 1$.

(II) If $k(\zeta_4)/k$ is unramified (including the case $\zeta_4 \in k$), then $S(k) = 1$.

We will give an example for each of the above cases: (I)-(i) $k = Q_2$. (I)-(ii) $k = Q_2(\sqrt{2}) \subset Q_2(\zeta_8)$. (I)-(iii) $k = Q_2(\sqrt{-2}) \subset Q_2(\zeta_8)$. (II) $k = Q_2(\sqrt{3}) \subset Q_2(\zeta_{12})$, where $k(\zeta_4) = Q_2(\zeta_{12})$ and $k(\zeta_4)/k$ is unramified of degree 2.

Proof of Theorem 5.11. Let B be an arbitrary cyclotomic algebra over k. We may assume that B is of the form:

$$B = (\beta, L/k), \quad L = Q_2(\zeta_{2^n}, \zeta_r) \tag{5.68}$$

$$2 \leqq n, \quad r = 2^a - 1, \tag{5.69}$$

where n, a are some integers. Then

$$L \supset M = k(\zeta_{2^h}) \supset k, \quad \text{so} \quad h \leqq n.$$

The inertia group T_0 of L/Q_2 is:

$$T_0 = \langle\theta\rangle \times \langle\iota\rangle, \quad \theta^{2^{n-2}} = \iota^2 = 1,$$

where θ and ι are defined by (5.8). Let T denote the inertia group of L/k. It follows from Lemma 5.10 that there are just four cases stated in the theorem. We have seen that T must be one of the following:

$$1) \ \langle\theta^{2^\lambda}\rangle \times \langle\iota\rangle, \quad 2) \ \langle\theta^{2^\lambda}\rangle, \quad 3) \ \langle\iota\theta^{2^\nu}\rangle,$$

where $0 \leqq \lambda \leqq n - 2$, $0 \leqq \nu \leqq n - 3$. First we will prove

<u>Lemma 5.12</u>. Let the notation be the same as in Theorem 5.11.

(I) Suppose that $k(\zeta_4)/k$ is ramified.

(i) If $h = 0$, then $T = \langle\theta\rangle \times \langle\iota\rangle$.

(ii) If $h \geqq 3$ and $z \equiv -1 \pmod{2^h}$, then $T = \langle\theta^{2^{h-2}}\rangle \times \langle\iota\rangle$.

(iii) If $h \geqq 3$ and $z \equiv -1 + 2^{h-1} \pmod{2^h}$, then $T = \langle\iota\theta^{2^{h-3}}\rangle$.

(II) If $k(\zeta_4)/k$ is unramified, then $T = \langle\theta^{2^{h-2}}\rangle$.

Proof. If $h = 0$, then k/Q_2 is unramified, and hence $T = \langle\theta\rangle \times \langle\iota\rangle$. For the rest of the proof, we assume $h \geqq 2$. The inertia group T' of $k(\zeta_4)/k$ is canonically isomorphic to $T\cdot H/H$, where $H = G(L/k(\zeta_4))$. Note that

$$\theta^{2^\lambda}(\zeta_4) = \zeta_4, \quad \iota(\zeta_4) = \iota\theta^{2^\nu}(\zeta_4) = \zeta_4^{-1}.$$

Hence $k(\zeta_4)/k$ is unramified, if and only if $T = \langle\theta^{2^\lambda}\rangle$ for some λ. In this case, $e_{M/k} = 1$ by Lemma 5.9 ($M = k(\zeta_{2^h})$). Because $e_{M/Q_2} = 2^{h-1}$ and $e_{L/Q_2} = 2^{n-1}$, it follows that $e_{L/k} = 2^{n-h}$, and hence $\lambda = h - 2$, $T = \langle\theta^{2^{h-2}}\rangle$, proving the assertion in (II).

Suppose that $k(\zeta_4)/k$ is ramified, i.e., either $T = \langle\theta^{2^\lambda}\rangle \times \langle\iota\rangle$ for some λ, or $T = \langle\iota\theta^{2^\nu}\rangle$ for some ν. Then by Lemma 5.10, $e_{M/k} = 2$, and so $e_{L/k} = 2\cdot 2^{n-h}$. Hence either $T = \langle\iota\rangle \times \langle\theta^{2^{h-2}}\rangle$, or $T = \langle\iota\theta^{2^{h-3}}\rangle$. We notice that

$$\iota(\zeta_{2^h}) = \zeta_{2^h}^{-1}, \quad \theta^{2^{h-2}}(\zeta_{2^h}) = \zeta_{2^h},$$

$$\iota\theta^{2^{h-3}}(\zeta_{2^h}) = \zeta_{2^h}^{z'}, \quad z' \equiv -5^{2^{h-3}} \equiv -1 + 2^{h-1} \pmod{2^h}.$$

From this the assertions in (I)-(ii), (I)-(iii) of the lemma follow immediately. #

We now return to the proof of Theorem 5.11. If $k(\zeta_4)/k$

is ramified, $h \geq 3$, and $z \equiv -1 + 2^{h-1} \pmod{2^h}$, then by Lemma 5.12, $T = \langle \iota\theta^{2^{h-3}} \rangle$. If $k(\zeta_4)/k$ is unramified, then by Lemma 5.12, $T = \langle \theta^{2^{h-2}} \rangle$. Theorem 5.3 now yields that for the above two cases, the cyclotomic algebra B given by (5.68) is similar to k: $B = (\beta, L/k) \sim k$. Since B is an arbitrary cyclotomic algebra over k, we conclude that for the cases (I)-(iii) and (II), $S(k) = 1$.

Now Corollary 5.5 implies that for the cases (I)-(i) and (I)-(ii), it suffices to construct a cyclotomic algebra over k whose invariant is $\frac{1}{2}$. Suppose that either $h = 0$, or $k(\zeta_4)/k$ is ramified, $h \geq 3$, $z \equiv -1 \pmod{2^h}$. Put $K = k(\zeta_{2^\ell})$, where $\ell = 2$ for $h = 0$, and $\ell = h$ for $h \geq 3$. By Lemma 5.8, $K = Q_2(\zeta_{2^\ell}, \zeta_{2^f-1})$, $f = f_{K/Q_2}$. (If $h = 0$, this is obvious.) Put

$$L = Q_2(\zeta_{2^\ell}, \zeta_{2^{2f}-1}).$$

Then L is the unramified extension of K of degree 2. Denote by E the maximal unramified extension of k in L. By Lemma 5.9, $L = k(\zeta_4)\cdot E$ and $k(\zeta_4) \cap E = k$. (If $h = 0$, this statement is obvious, because $E = Q_2(\zeta_{2^{2f}-1})$ and $k = Q_2(\zeta_{2^f-1})$.) Let E' be the maximal unramified extension of k in K ($E' = k$ for $h = 0$). Then $G(L/E)$ is canonically isomorphic to $G(K/E')$, so that for the cases (I)-(i) and (I)-(ii), we have

$$G(L/E) = \langle\omega\rangle, \quad \omega^2 = 1, \quad \zeta_{2^\ell}^{\omega} = \zeta_{2^\ell}^{-1}. \tag{5.70}$$

Let ϕ be a generating automorphism of the unramified extension $L/k(\zeta_4)$. Then $G(L/k) = \langle\omega\rangle \times \langle\phi\rangle$. Set $r = [K : k(\zeta_4)]$, so that

$$[L : k(\zeta_4)] = [E : k] = 2r \quad \text{and} \quad \phi^{2r} = 1.$$

For convenience sake, we choose the odd numbers $3, 5, \cdots, 2^\ell - 1, 2^\ell + 1$ as a complete system of representatives of integers mod 2^ℓ relatively prime to 2. Let ψ denote the restriction of ϕ to K. Then $G(K/k(\zeta_4)) = \langle\psi\rangle$, $\psi^r = 1$, and

$$\zeta_{2^\ell}^{\psi} = \zeta_{2^\ell}^{\phi} = \zeta_{2^\ell}^{t} \tag{5.71}$$

for a certain integer t such that $(2, t) = 1$ and $3 \leqq t \leqq 2^\ell + 1$. Note that $t = 1 + 2^2$ for $h = 0$, because in this case, $K = k(\zeta_4)$ and $\psi = 1$. Write

$$t = 1 + 2^a m, \quad (2, m) = 1, \quad 2 \leqq 2^a m \leqq 2^\ell.$$

Since $\zeta_4^t = \zeta_4^\psi = \zeta_4$, it follows that $2 \leqq a \leqq \ell$. Consequently, the order of t mod 2^ℓ in Z mod$^\times 2^\ell$ equals $2^{\ell-a}$. (For $\ell \geqq 3$, this fact is wellknown and follows easily, for instance, from [46, Lemma 1]. For $\ell = 2$, the statement is evident, because $t = 1 + 2^2$, $a = 2$.) Hence the order of ψ, which is equal to

r, is divisible by $2^{\ell-a}$. We easily conclude from this that the number $t^{2r} - 1 = (1 + 2^a m)^{2r} - 1$ is divisible by $2^{\ell+1}m$. Put

$$y = (t^{2r} - 1)/2^{\ell+1}m. \tag{5.72}$$

We will construct a cyclotomic algebra B of L/k as follows: Set

$$h_\phi = \zeta_{2^a}^{-y}, \quad h_\omega = 1, \quad h_{\omega,\phi} = \zeta_{2^\ell}, \quad h_{\phi,\omega} = \zeta_{2^\ell}^{-1}, \quad (\zeta_{2^\ell}^{2^{\ell-a}} = \zeta_{2^a}).$$

Then $\zeta_{2^a}^\phi = \zeta_{2^a}^t = \zeta_{2^a}$, and so $(h_\phi)^\phi = h_\phi$. Obviously, $(h_\omega)^\omega = h_\omega$. We have $h_\omega^{\phi-1} = 1$ and $h_{\phi,\omega}^{1+\omega} = \zeta_{2^\ell}^{-(1-1)} = 1$, so that $h_\omega^{\phi-1} = h_{\phi,\omega}^{1+\omega}$. Because $1 + t + \cdots + t^{2r-1} = (t^{2r} - 1)/(t - 1) = y2^{\ell+1-a}$, it follows that

$$h_{\omega,\phi}^{1+\phi+\cdots+\phi^{2r-1}} = \zeta_{2^\ell}^{1+t+\cdots+t^{2r-1}} = \zeta_{2^a}^{2y}.$$

On the other hand, we have $h_\phi^{\omega-1} = (\zeta_{2^a}^{-y})^{-1-1} = \zeta_{2^a}^{2y}$. Consequently, $h_\phi^{\omega-1} = h_{\omega,\phi}^{1+\phi+\cdots+\phi^{2r-1}}$. Thus the elements h_ϕ, h_ω, $h_{\omega,\phi}$, $h_{\phi,\omega}$ satisfy the relations (2.16)-(2.18), and so give rise to a cyclotomic algebra B over k (cf. Chapter 2):

$$B = (\beta, L/k) = \sum_{i=0}^{1} \sum_{j=0}^{2r-1} Lu_\omega^i u_\phi^j,$$

$$u_\omega u_\phi = \zeta_{2^\ell} u_\phi u_\omega, \quad u_\omega^2 = 1, \quad u_\phi^{2r} = \zeta_{2^a}^{-y},$$

$$u_\omega^i u_\phi^j x = \omega^i \phi^j (x) \cdot u_\omega^i u_\phi^j \quad (x \in L),$$

$$(u_\omega^i u_\phi^j)(u_\omega^{i'} u_\phi^{j'}) = \beta(\omega^i \phi^j, \omega^{i'} \phi^{j'}) u_\omega^{|i+i'|_2} u_\phi^{|j+j'|_{2r}},$$

$$0 \leqq i, i' \leqq 1, \quad 0 \leqq j, j' \leqq 2r - 1,$$

where for a natural number n and an integer c, $|c|_n$ denotes the integer s such that $s \equiv c \pmod{n}$, $0 \leqq s \leqq n - 1$. We have

$$\beta(\omega, \phi)/\beta(\phi, \omega) = u_\omega u_\phi u_\omega^{-1} u_\phi^{-1} = \zeta_{2^\ell}, \quad \beta(\omega, \omega) = u_\omega^2 = 1.$$

Then Theorem 5.6 yields that

$$\mathrm{inv}_k(B) \equiv \frac{1}{2} \pmod{Z}.$$

This completes the proof of Theorem 5.11.

<u>Remark 5.13</u>. Recall that the residue class degree of the extension L/k is $2r$. Indeed, it can be proved that if the residue class degree of $Q_2(\zeta)/k$ is odd, ζ being a root of unity, then any cyclotomic algebra of the form $(\beta, Q_2(\zeta)/k)$ is similar to k, except the case $\iota \in G(Q_2(\zeta)/k)$, $\beta(\iota, \iota) = -1$, $2 \nmid [k : Q_2]$.

Let k be a cyclotomic extension of Q_2 with the height

h, and let $L = Q_2(\zeta_{2^n}, \zeta_r)$, $2 \leqq n$, $(2, r) = 1$, be a cyclotomic field containing k (cf. (5.68), (5.69)). Then $h \leqq n$. Let T denote the inertia group of L/k. We conclude from Lemma 5.12 that for the cases (I)-(iii) and (II) T is cyclic, and that for the cases (I)-(i) and (I)-(ii), T is not cyclic provided that $n > \mathrm{Max}(h, 2)$. Hence we have the following, which is equivalent to Theorem 5.11:

Theorem 5.14 (Yamada [54]). Let k be a cyclotomic extension of Q_2. If there exists a root of unity ζ such that $Q_2(\zeta) \supset k$ and that the inertia group of the extension $Q_2(\zeta)/k$ is not cyclic, then $S(k)$ is the subgroup of $Br(k)$ of order 2. Otherwise, $S(k) = 1$. (That is, if the inertia group of $Q_2(\zeta)/k$ is cyclic for any arbitrary root of unity ζ such that $Q_2(\zeta) \supset k$, then $S(k) = 1$.)

Finally we state

Theorem 5.15 (Yamada [51]). Let K be a finite extension of Q_2. Let k be the maximal cyclotomic extension of Q_2 contained in K.

1) If $[K : k]$ is even, then $S(K) = 1$.

2) If $[K : k]$ is odd, then $S(K) = S(k) \otimes_k K \cong S(k)$, and $S(k)$ is determined by Theorem 5.11.

Proof. This follows at once from Proposition 4.6.

Chapter 6. PROPERTIES OF A SCHUR ALGEBRA

The purpose of this chapter is to prove

Theorem 6.1 (Benard-Schacher). Let k be a cyclotomic extension of Q, the rationals. Let $[A] \in S(k)$, and suppose A has index m. Then a primitive m-th root of unity ζ_m belongs to k: $\zeta_m \in k$. Let p be a rational prime. If $\mathfrak{p}$ is a prime of k lying above p and $\theta \in G(k/Q)$ with $\zeta_m^\theta = \zeta_m^b$, b being an integer with $(b, m) = 1$, then

$$\mathrm{inv}_{\mathfrak{p}}(A) \equiv b \cdot \mathrm{inv}_{\mathfrak{p}^\theta}(A) \pmod{Z}. \tag{6.1}$$

A simple proof of the first assertion in the theorem was given by Janusz [31]:

Proposition 6.2. Let K be a field. If a cyclotomic algebra B over K has exponent m in the Brauer group $Br(K)$ of K, then the m-th roots of unity lie in the center K.

Proof. Let

$$B = (\beta, K(\zeta)/K) = \sum_{\sigma \in G} K(\zeta)u_\sigma,$$

$$u_\sigma x = x^\sigma u_\sigma \quad (x \in K(\zeta)), \qquad u_\sigma u_\tau = \beta(\sigma, \tau)u_{\sigma\tau}$$

be a cyclotomic algebra over K, where ζ is a root of unity,

$G = G(K(\zeta)/K)$, and β is a factor set whose values are roots of unity in $K(\zeta)$. We shall change factor sets in the proof so we write $B(\beta)$ for the crossed product made with $K(\zeta)$, G and β. Let the values of β generate a group $\langle\zeta_n\rangle$ of n-th roots of unity. In the Brauer group of K, we have

$$[B(\beta)]^n = [B(\beta^n)] = [B(1)] = 1.$$

By assumption the order of $[B(\beta)]$ is m so m divides n. Thus a primitive m-th root ζ_m lies in $K(\zeta)$ and it is necessary to show it lies in K. We show this by proving ζ_m is fixed by all elements in G. Let $\tau \in G$ and let $B(\beta^\tau)$ denote the crossed product

$$(\beta^\tau, K(\zeta)/K) = \sum_{\sigma\in G} K(\zeta)v_\sigma,$$

where the new factor set β^τ is obtained by applying τ to the factor set β. It is necessary to check that β^τ is indeed a factor set. This is not in general true but works here because G is abelian. The verification is left to the reader. Now map $B(\beta)$ to $B(\beta^\tau)$ by $\Sigma\, x_\sigma u_\sigma \to \Sigma\, \tau(x_\sigma)v_\sigma$. One may now check this is a K-algebra isomorphism and so $[B(\beta)] = [B(\beta^\tau)]$ in Br(K). Now there is a positive integer r such that $\tau(\zeta_n) = \zeta_n^r$. It follows that $\beta^\tau = \beta^r$. Thus

$$[B(\beta)] = [B(\beta^r)] = [B(\beta)]^r.$$

Since $[B(\beta)]$ has order m we find m divides $r - 1$; or $r = 1 + ms$, $s \in Z$. Now ζ_m lies in $<\zeta_n>$ so $\tau(\zeta_m) = \zeta_m^r = \zeta_m^{1+ms} = \zeta_m$ which proves ζ_m is fixed by all elements of G as required. #

Proof of Theorem 6.1. In view of Corollary 3.11, we need only prove the second assertion for a cyclotomic algebra over k. Let

$$B = (\beta,\ k(\zeta)/k) = \sum_{\sigma\varepsilon G} k(\zeta)u_\sigma, \qquad u_\sigma u_\tau = \beta(\sigma,\ \tau)u_{\sigma\tau}$$

be a cyclotomic algebra over k. The automorphism θ of the extension k/Q can be extended to an automorphism of $k(\zeta)/Q$, which is also denoted by θ. Put

$$\beta^\theta(\sigma,\ \tau) = (\beta(\sigma,\ \tau))^\theta \quad \text{for} \quad \sigma,\ \tau \varepsilon\ G = G(k(\zeta)/k).$$

Then β^θ is also a factor set of $k(\zeta)/k$, because $k(\zeta)/Q$ is an abelian extension. Let B^θ denote the cyclotomic algebra with the factor set β^θ:

$$B^\theta = (\beta^\theta,\ k(\zeta)/k) = \sum_{\sigma\varepsilon G} k(\zeta)v_\sigma, \qquad v_\sigma v_\tau = \beta^\theta(\sigma,\ \tau)v_{\sigma\tau}.$$

Now map B onto B^θ by

$$\sum_{\sigma\varepsilon G} x_\sigma u_\sigma \to \sum_{\sigma\varepsilon G} x_\sigma^\theta v_\sigma, \qquad x_\sigma \in k(\zeta).$$

One may now check this is a Q-algebra isomorphism and $\mathrm{inv}_p(B) =$

$\mathrm{inv}_{p^\theta}(B^\theta)$. Let the roots of unity in $k(\zeta)$ generate a cyclic group $\langle\zeta_n\rangle$ of n-th roots of unity. Then $m|n$. If $\zeta_n^\theta = \zeta_n^t$ for an integer t, then $\zeta_m^\theta = \zeta_m^t$, and hence $t \equiv b \pmod m$. We thus conclude $[B^\theta] = [B]^t$ in $Br(k)$ and

$$\mathrm{inv}_p(B) = \mathrm{inv}_{p^\theta}(B^\theta) \equiv t\cdot\mathrm{inv}_{p^\theta}(B) \equiv b\cdot\mathrm{inv}_{p^\theta}(B) \pmod{Z}.$$

The proof of Theorem 6.1 is completed.

<u>Corollary 6.3</u> (Benard [4]). Let k be a cyclotomic extension of Q and $[A] \in S(k)$. Let p be a rational prime. If p and p' are primes of k lying above p, then $A \otimes_k k_p$ and $A \otimes_k k_{p'}$ have the same index.

<u>Proof</u>. This is clear from (6.1), because $(b, m) = 1$ and the denominator of $\mathrm{inv}_p(A)$ divides m.

<u>Definition 6.4</u>. Let k be a cyclotomic extension of Q and $[A] \in S(k)$. Let p be a rational prime. Then the common value of the indices of $A \otimes_k k_p$ for all primes p of k lying above p is called <u>the p-local index of</u> A. For a prime p lying above p, the invariant of $A \otimes_k k_p$, $\mathrm{inv}_p(A)$, is called <u>a p-local invariant of</u> A. If it is equal to $\frac{1}{2}$, then we may call it <u>the p-local invariant of</u> A, because $\mathrm{inv}_p(A) = \mathrm{inv}_{p'}(A) = \frac{1}{2}$ for any p, p' dividing p. In particular, if the index of A is 1 or 2, then we can speak of the p-local invariant of A

for any rational prime p.

Definition 6.5. Let k be a cyclotomic extension of Q. If a central simple algebra A over k has invariants which are distributed in the way described in Theorem 6.1 for all rational primes p, A is said to have uniformly distributed invariants.

We will state other corollaries of Theorem 6.1. The notation will be the same as in it.

Corollary 6.6. Let $[A] \in S(k)$ and suppose A has p-local index m'. Then each of the values $\frac{t}{m'}$, where $0 < t < m'$ and $(t, m') = 1$, occurs equally often as p-local invariants of A.

Proof. Note that the p-local index m' of A divides the index m of A. Hence if $\theta \in G(k/Q)$ and $\zeta_m^\theta = \zeta_m^b$, $(b, m) = 1$, then $\zeta_{m'}^\theta = \zeta_{m'}^b$. As θ ranges over all elements of $G(k/Q)$, b mod m' takes each element of $Z \bmod^{\times} m'$ an equal number of times. Since the denominator of $\mathrm{inv}_{\mathcal{p}}(A)$ is m', the equation (6.1) yields that each value $\frac{t}{m'}$ occurs equally often as $\mathrm{inv}_{\mathcal{p}^\theta}(A)$. #

Corollary 6.7. Let $[A] \in S(k)$ and let $\mathcal{p}$ and $\mathcal{p}'$ be primes of k dividing p. Suppose A has p-local index m'. Then

$$\mathrm{inv}_{p}(A) = \mathrm{inv}_{p'}(A) \Longleftrightarrow p \cap Q(\zeta_{m'}) = p' \cap Q(\zeta_{m'}).$$

Proof. If $\theta \in G(k/Q)$ with $\zeta_m^{\theta} = \zeta_m^{b}$ and so $\zeta_{m'}^{\theta} = \zeta_{m'}^{b}$, then $b \equiv 1 \pmod{m'}$ if and only if θ fixes $Q(\zeta_{m'})$ pointwise. By Theorem 4.3, $p \equiv 1 \pmod{m'}$, so p splits completely in $Q(\zeta_{m'})$. Thus θ fixes $Q(\zeta_{m'})$ pointwise if and only if $p \cap Q(\zeta_{m'}) = p^{\theta} \cap Q(\zeta_{m'})$. #

Theorem 6.1 implies the following.

Theorem 6.8. Let χ be an irreducible character. If $m = m_Q(\chi)$, then $\zeta_m \in Q(\chi)$.

We notice that the Brauer-Speiser theorem (Corollary 1.8) is also a corollary of Theorem 6.8.

Chapter 7. THE SCHUR SUBGROUP OF A REAL FIELD

Let k be a cyclotomic extension of the rational field Q. Throughout this chapter, k is assumed to be real.

Definition 7.1. The subgroup of Br(k) consisting of those algebra classes which have uniformly distributed invariants with values 0 or $\frac{1}{2}$ is denoted by M(k). If p is a rational prime and $[A] \in M(k)$, then we write $inv_p(A)$ to denote the p-local invariant of A (cf. Definition 6.4).

A class [A] of Br(k) belongs to M(k) if and only if the index of A is at most 2 and $inv_{\mathfrak{p}}(A) = inv_{\mathfrak{p}'}(A)$ for any primes $\mathfrak{p}$, $\mathfrak{p}'$ of k lying above an arbitrary rational prime p. It follows from the Brauer-Speiser theorem (Corollary 1.8) and the Benard-Schacher theorem (Theorem 6.1) that the Schur subgroup S(k) is contained in M(k). On the other hand, S(k) contains the subgroup

$$S(Q) \otimes_Q k = \{[A' \otimes_Q k];\ \ [A'] \in S(Q)\}, \tag{7.1}$$

whose elements are induced from those of S(Q), the Schur subgroup of the rationals. Thus we have

$$Br(k) \supset M(k) \supset S(k) \supset S(Q) \otimes_Q k. \tag{7.2}$$

First we will prove:

Theorem 7.2 (Benard [2] and Fields [22]). The Schur subgroup $S(Q)$ of the rationals is the subgroup $M(Q)$ of $Br(Q)$: $S(Q) = M(Q)$. That is, an algebra class $[A]$ of $Br(Q)$ belongs to $S(Q)$ if and only if the index of A is either 1 or 2.

Proof. Let p be an odd prime and ζ_p a primitive p-th root of unity. Consider the following cyclic algebra B_p over Q:

$$B_p = (-1, Q(\zeta_p)/Q, \sigma), \quad \langle\sigma\rangle = G(Q(\zeta_p)/Q), \tag{7.3}$$

which is also a cyclotomic algebra over Q. For any finite rational prime $p' \neq p$, we have $\mathrm{inv}_{p'}(B_p) = 0$, because p' is unramified in $Q(\zeta_p)/Q$ and -1 is a unit. On the other hand, using Theorem 4.3 we conclude $\mathrm{inv}_p(B_p) = \frac{1}{2}$. (See also Appendix to Chapter 4.) Then Hasse's sum theorem yields $\mathrm{inv}_\infty(B_p) = \frac{1}{2}$, where ∞ denotes the infinite prime.

Consider next the following quaternion algebra B_2 over Q:

$$B_2 = (-1, Q(\zeta_4)/Q, \sigma), \quad \zeta_4^\sigma = \zeta_4^{-1}, \tag{7.4}$$

which is also a cyclotomic algebra over Q. For any finite rational prime $p' \neq 2$, $\mathrm{inv}_{p'}(B_2) = 0$. Using the formula (5.40) of Theorem 5.6 we conclude $\mathrm{inv}_2(B_2) = \frac{1}{2}$, so by Hasse's sum theorem, $\mathrm{inv}_\infty(B_2) = \frac{1}{2}$. By taking tensor products of above cyclic algebras, we see that every algebra class represented by a quaternion division algebra central over Q belongs to $S(Q)$.

Theorem 7.3 (Yamada [48]). Let k be a cyclotomic extension of Q such that $[k : Q]$ is odd. Then k is real and $M(k) = S(k) = S(Q) \otimes_Q k$. That is, $S(k)$ consists of the classes of $Br(k)$ which have uniformly distributed invariants with values 0 or $\frac{1}{2}$.

Proof. The first assertion is clear. Let $[A] \in M(k)$. Let $\{p_1, p_2, \cdots, p_s\}$ be the set of rational primes with $\mathrm{inv}_{p_i}(A) = \frac{1}{2}$. For each i, the number of the primes of k dividing p_i is odd, because $[k : Q]$ is odd. By Hasse's sum theorem, s must be even. Let D denote the quaternion algebra central over Q with invariant $\frac{1}{2}$ at p_i $(i = 1, 2, \cdots, s)$ and 0 elsewhere. Since $[k : Q]$ is odd, the local degree $[k_{\mathfrak{p}} : Q_p]$ is odd for any prime $\mathfrak{p}$ of k dividing a rational prime p. Hence we have $[D \otimes_Q k] = [A]$. By Theorem 7.2, $[D] \in S(Q)$. We thus conclude $M(k) = S(k) = S(Q) \otimes_Q k$.

By making use of the results of Chapter 2 about cyclotomic algebras, and the formula of index of a p-adic cyclotomic algebra, we can determine the Schur subgroup of a real subfield of the cyclotomic field $Q(\zeta_{\ell^n})$:

Theorem 7.4 (Yamada [48]). Let ℓ be a prime number, n a positive integer, and ζ_{ℓ^n} a primitive ℓ^n-th root of unity. Let k be a real subfield of the cyclotomic field $Q(\zeta_{\ell^n})$. Then, $S(k) = M(k)$. That is, $S(k)$ consists of the classes of

Br(k) which have uniformly distributed invariants with values 0 or $\frac{1}{2}$.

<u>Proof</u>. By Theorem 7.3 we may assume that $[k : Q]$ is even. In particular we have done with the case $\ell \equiv 3 \pmod 4$, because $[k : Q]$ divides $\ell^{n-1}(\ell - 1)/2 = [Q(\zeta_{\ell^n} + \zeta_{\ell^n}^{-1}) : Q]$, which is odd for $\ell \equiv 3 \pmod 4$.

For the rest of the proof we assume that either $\ell \equiv 1 \pmod 4$ or $\ell = 2$, and that $[k : Q]$ is even. Recall that the prime ℓ is totally ramified in k/Q. Let $\mathfrak{l}$ be the unique prime of k dividing ℓ. For each rational prime p other than ℓ, g_p denotes the number of the primes $\mathfrak{p}$ of k lying above p. If g_p is even then $\Omega(p)$ denotes the class of Br(k) with invariant $\frac{1}{2}$ at the primes $\mathfrak{p}$ and 0 elsewhere. If g_p is odd then $\Omega(\ell, p)$ denotes the class of Br(k) with invariant $\frac{1}{2}$ at the prime $\mathfrak{l}$ and the primes $\mathfrak{p}$ dividing p and 0 elsewhere. Since k/Q is cyclic, there are infinitely many p with odd g_p. It is easily verified that if for each p with even g_p (resp. odd g_p) the above $\Omega(p)$ (resp. $\Omega(\ell, p)$) belongs to S(k), then S(k) consists of the classes which have uniformly distributed invariants with values 0 or $\frac{1}{2}$.

Let f_p denote the residue class degree of p ($\neq \ell$) in k/Q. If f_p is odd then g_p is even, for $[k : Q] = f_p g_p$ is even. In this case, $\Omega(p)$ is in S(k), because $\Omega(p) = [D_{\ell,p} \otimes_Q k]$, where $D_{\ell,p}$ denotes the quaternion division algebra central over Q with invariant $\frac{1}{2}$ at ℓ and p and 0 elsewhere. In particular, $\Omega(\infty)$ is in S(k), where $p = \infty$ is the rational infinite prime. For each p with even f_p, we will construct a cyclotomic algebra B_p over k with invariant $\frac{1}{2}$ at the primes $\mathfrak{p}$ of k dividing p and 0 at any finite prime of k dividing neither p nor ℓ. Then it follows from Hasse's sum theorem that if g_p is even (resp. odd), B_p has invariant 0 (resp. $\frac{1}{2}$) at the prime $\mathcal{l}$, because g_∞ is even and a cyclotomic algebra has the same invariant at the infinite primes. Since $\Omega(\infty)$ is in S(k), we conclude that if g_p is even (resp. odd) then $\Omega(p)$ (resp. $\Omega(\ell, p)$) is in S(k), and so the theorem will be proved.

Suppose that $\ell \equiv 1 \pmod 4$. Let p be a prime number with even f_p. Let $\mathfrak{p}$ be any prime of k dividing p. Denote by g' the number of the primes of $Q(\zeta_{\ell^n})$ dividing $\mathfrak{p}$. Denote by f' the residue class degree of $\mathfrak{p}$ in $Q(\zeta_{\ell^n})/k$. Then g' must be odd. For, suppose that g' would be even. Since the extension $Q(\zeta_{\ell^n})/Q$ is cyclic and the residue class degree of p in $Q(\zeta_{\ell^n})/Q$ is f_pf', which is divisible by 2f', the unique cyclic subgroup H of $G(Q(\zeta_{\ell^n})/Q)$ of order 2f' is contained in the decomposition group of p in $Q(\zeta_{\ell^n})/Q$. But H is also contained in $G(Q(\zeta_{\ell^n})/k)$, because $f'g' = [Q(\zeta_{\ell^n}) : k]$ is divisible by 2f'. Consequently the residue class degree f' of $\mathfrak{p}$ in $Q(\zeta_{\ell^n})/k$ is divisible by 2f'. This is a contradiction. Hence g' must be odd.

First consider an odd prime p with even f_p. Set $K = Q(\zeta_{\ell^n}, \zeta_p)$. Then $K = Q(\zeta_{\ell^n})\cdot k(\zeta_p)$ and $Q(\zeta_{\ell^n}) \cap k(\zeta_p) = k$. The extensions $K/Q(\zeta_{\ell^n})$ and $K/k(\zeta_p)$ are cyclic. Let θ (resp. ϕ) be a generating automorphism of $K/Q(\zeta_{\ell^n})$ (resp. $K/k(\zeta_p)$). Then $G = G(K/k) = \langle\theta\rangle \times \langle\phi\rangle$. Set $[K : k(\zeta_p)] = [Q(\zeta_{\ell^n}) : k] = s$. Since k is a real subfield of $Q(\zeta_{\ell^n})$, s is even. Set $h_{\theta,\phi} = h_{\phi,\theta} = -1$ and $h_\theta = h_\phi = 1$. As is easily seen, these elements satisfy the relations (2.16)-(2.19). (The relation (2.19) does not matter in this case, for G is a direct product of two cyclic groups.) Hence we get a cyclotomic algebra B_p over k:

$$B_p = (\beta, K/k) = \sum_{i=0}^{p-2} \sum_{j=0}^{s-1} K u_\theta^i u_\phi^j \qquad \text{(direct sum)},$$

$$u_\theta^i u_\phi^j \cdot x = x^{\theta^i \phi^j} \cdot u_\theta^i u_\phi^j, \qquad (x \in K),$$

$$u_\theta u_\phi = -u_\phi u_\theta, \qquad u_\theta^{p-1} = u_\phi^s = 1,$$

$$(u_\theta^i u_\phi^j)(u_\theta^{i'} u_\phi^{j'}) = \beta(\theta^i\phi^j, \theta^{i'}\phi^{j'}) u_\theta^{i''} u_\phi^{j''},$$

$$0 \leqq i, i', i'' \leqq p - 2, \quad 0 \leqq j, j', j'' \leqq s - 1,$$

$$i + i' \equiv i'' \pmod{p - 1}, \quad j + j' \equiv j'' \pmod{s}.$$

B_p has invariant 0 at any finite prime $\mathfrak{q}$ of k dividing neither p nor ℓ, because $\mathfrak{q}$ is unramified in K/k.

Let $\mathfrak{p}$ be any prime of k dividing p. Since p is unramified in $Q(\zeta_{\ell^n})/Q$ and totally ramified in $K/Q(\zeta_{\ell^n})$, we see easily that $G(K^{p}/k_{p}) = \langle\theta\rangle \times \langle\phi^{g'}\rangle$ and that for some integer a, $0 \leq a < s$, ϕ^{a} is a Frobenius automorphism of p in K/k with $\langle\phi^{a}\rangle = \langle\phi^{g'}\rangle$. Since g' is odd and the order of ϕ is s, which is even, it follows that a is also odd, and so

$$u_{\theta}u_{\phi}^{a} = -u_{\phi}^{a}u_{\theta}, \qquad \beta(\theta, \phi^{a})/\beta(\phi^{a}, \theta) = -1.$$

Set $q = N_{k/Q}(p)$. Then

$$\begin{aligned}\delta &= (\beta(\theta,\phi^{a})/\beta(\phi^{a},\theta))^{(p-1)/(q-1)}\beta(\theta,\theta)\beta(\theta^{2},\theta)\cdots\beta(\theta^{p-2},\theta)\\ &= (-1)^{(p-1)/(q-1)}u_{\theta}^{p-1} = \zeta_{q-1}^{(p-1)/2}.\end{aligned}$$

Hence from the formula (4.9) of Theorem 4.3, we conclude that the p-index of the cyclotomic algebra B_{p} equals 2, as required.

Next suppose that the prime 2 has even f_{2}. Set $K = Q(\zeta_{\ell^n}, \zeta_{4})$. Then $K = Q(\zeta_{\ell^n})k(\zeta_{4})$ and $Q(\zeta_{\ell^n}) \cap k(\zeta_{4}) = k$. The extensions $K/Q(\zeta_{\ell^n})$ and $K/k(\zeta_{4})$ are cyclic. Let $G(K/Q(\zeta_{\ell^n})) = \langle\iota\rangle$ and $G(K/k(\zeta_{4})) = \langle\phi\rangle$. Then $\zeta_{4}^{\iota} = \zeta_{4}^{-1}$, $G(K/k) = \langle\iota\rangle \times \langle\phi\rangle$, $[K : k(\zeta_{4})] = [Q(\zeta_{\ell^n}) : k] = s$, $2|s$ and $\iota^{2} = \phi^{s} = 1$. Set $h_{\iota,\phi} = \zeta_{4}^{-1}$, $h_{\phi,\iota} = \zeta_{4}$, $h_{\iota} = 1$, and $h_{\phi} = \zeta_{4}^{s/2}$. Then these elements satisfy the relations (2.16)-(2.19) and so give rise to a cyclotomic algebra B_{2} over k:

$$B_{2} = (\beta, K/k) = \sum_{i=0}^{1}\sum_{j=0}^{s-1} Ku_{\iota}^{i}u_{\phi}^{j} \qquad \text{(direct sum)},$$

$$u_\phi u_\iota = \zeta_4 u_\iota u_\phi, \quad u_\iota^2 = 1, \quad u_\phi^s = \zeta_4^{s/2}. \tag{7.5}$$

B_2 has invariant 0 at any finite prime $\mathfrak{q}$ of k dividing neither 2 nor ℓ, because $\mathfrak{q}$ is unramified in K/k.

Let $\mathfrak{p}$ be any prime of k dividing 2. Since 2 is unramified in $Q(\zeta_{\ell^n})/Q$ and totally ramified in $K/Q(\zeta_{\ell^n})$, we see easily that $G(K^{\mathfrak{p}}/k_{\mathfrak{p}}) = \langle\iota\rangle \times \langle\phi^{g'}\rangle$ and that for some integer a, $0 \leq a < s$, ϕ^a is a Frobenius automorphism of $\mathfrak{p}$ in K/k with $\langle\phi^a\rangle = \langle\phi^{g'}\rangle$. By (7.5), we have

$$u_\iota u_\phi^a = \zeta_4^{-a}\, u_\phi^a u_\iota, \quad \beta(\iota, \phi^a)/\beta(\phi^a, \iota) = \zeta_4^{-a}. \tag{7.6}$$

Since g' is odd and the order of ϕ is s, which is even, it follows that a is also odd. Thus we see that the number b in (5.38) of Theorem 5.6 is, for our case, equal to a, and $2\nmid a$. Since $\beta(\iota, \iota) = u_\iota^2 = 1$, the formula (5.39) of Theorem 5.6 yields that

$$\mathrm{inv}_2(B_2) = \mathrm{inv}_{\mathfrak{p}}(B_2) \equiv \frac{a}{2} \equiv \frac{1}{2} \pmod{Z}.$$

This proves the theorem for the case $\ell \equiv 1 \pmod 4$.

Before proving the theorem for the case $\ell = 2$, we will determine the subfields of the cyclotomic field $Q(\zeta_{2^n})$, $n \geq 3$. The Galois group $G(Q(\zeta_{2^n})/Q)$, $n \geq 3$, is isomorphic to the multiplicative group of integers modulo 2^n, $Z \bmod^{\times} 2^n$, which is the direct product $\langle 5 \bmod 2^n\rangle \times \langle -1 \bmod 2^n\rangle$ of two cyclic groups of order 2^{n-2} and order 2. Hence we have

$$G(Q(\zeta_{2^n})/Q) = \langle\theta\rangle \times \langle\iota\rangle, \quad \theta^{2^{n-2}} = \iota^2 = 1, \tag{7.7}$$

$$\zeta_{2^n}^{\theta} = \zeta_{2^n}^{5}, \qquad \zeta_{2^n}^{\iota} = \zeta_{2^n}^{-1}. \tag{7.8}$$

It is easy to see that the subgroups of $\langle\theta\rangle \times \langle\iota\rangle$ are classified into three types:

$$\text{(i)} \quad \langle\theta^{2^\lambda}\rangle \times \langle\iota\rangle, \quad (\lambda = 0, 1, \cdots, n-2), \tag{7.9}$$

$$\text{(ii)} \quad \langle\theta^{2^\lambda}\rangle, \quad (\lambda = 0, 1, \cdots, n-2), \tag{7.10}$$

$$\text{(iii)} \quad \langle\iota\theta^{2^\nu}\rangle, \quad (\nu = 0, 1, \cdots, n-3). \tag{7.11}$$

The fixed fields of these subgroups are, respectively,

$$\text{(i)} \quad Q(\zeta_{2^{\lambda+2}} + \zeta_{2^{\lambda+2}}^{-1}), \quad (\lambda = 0,1,\cdots,n-2), \tag{7.12}$$

$$\text{(ii)} \quad Q(\zeta_{2^{\lambda+2}}), \quad (\lambda = 0,1,\cdots,n-2), \tag{7.13}$$

$$\text{(iii)} \quad Q(\zeta_4\cdot(\zeta_{2^{\nu+3}} + \zeta_{2^{\nu+3}}^{-1})), \quad (\nu = 0,1,\cdots,n-3). \tag{7.14}$$

The case (iii) is not so obvious:

Proposition 7.5. Let the notation be as above. The fixed field of the subgroup $\langle\iota\theta^{2^\nu}\rangle$ of $G = G(Q(\zeta_{2^n})/Q)$ is the field

$$Q(\zeta_4 \cdot (\zeta_{2^{\nu+3}} + \zeta_{2^{\nu+3}}^{-1})), \qquad (\nu = 0,1,\cdots,n-3), \qquad (7.15)$$

which is a cyclic extension of Q and is not real. The roots of unity contained in it are ± 1.

<u>Proof.</u> Put $H = \langle \iota\theta^{2^\nu} \rangle$ and denote by E_ν the fixed field of H in $Q(\zeta_{2^n})$. We notice that $\theta^{2^{\nu+1}} \in H$ and $\theta^{2^\nu} \notin H$. Hence the order of the factor group G/H is at least $2^{\nu+1}$. On the other hand, $|H| = 2^{n-2-\nu}$, and $2^{\nu+1} \cdot |H| = 2^{\nu+1} \cdot 2^{n-2-\nu} = 2^{n-1} = |G|$. We thus conclude that the factor group G/H is cyclic of order $2^{\nu+1}$, and hence E_ν is cyclic over Q with $[E_\nu : Q] = 2^{\nu+1}$. Since $\langle \iota\theta^{2^\nu} \rangle$ is not a subgroup of $\langle\theta\rangle$ and $Q(\zeta_4)$ is the fixed field of $\langle\theta\rangle$, it follows that $\zeta_4 \notin E_\nu$ and the roots of unity in E_ν are ± 1. Denote by E'_ν the field given by (7.15). Since ζ_4 is not real and $\zeta_{2^{\nu+3}} + \zeta_{2^{\nu+3}}^{-1}$ is real, the field E'_ν is not real. By Lemma 5.2, we have $5^{2^\nu} = 1 + 2^{\nu+2}\kappa$, $(2, \kappa) = 1$. Hence

$$\iota\theta^{2^\nu}(\zeta_4 \cdot (\zeta_{2^{\nu+3}} + \zeta_{2^{\nu+3}}^{-1})) = \zeta_4^{-1} \cdot (\zeta_{2^{\nu+3}}^{-(1+2^{\nu+2}\kappa)} + \zeta_{2^{\nu+3}}^{(1+2^{\nu+2}\kappa)})$$

$$= -\zeta_4 \cdot (-\zeta_{2^{\nu+3}}^{-1} - \zeta_{2^{\nu+3}}) = \zeta_4 \cdot (\zeta_{2^{\nu+3}} + \zeta_{2^{\nu+3}}^{-1}),$$

$$(\zeta_{2^{\nu+3}}^{\pm 2^{\nu+2}\kappa} = (-1)^{\pm\kappa} = -1).$$

Thus E'_ν is contained in E_ν. Since the extension E_ν/Q is cyclic of degree $2^{\nu+1}$, the fields between E_ν and Q are linearly ordered. The field $Q(\zeta_{2^{\nu+2}} + \zeta_{2^{\nu+2}}^{-1})$ is the maximal subfield of E_ν over Q, because it has degree 2^ν over Q and is contained in E_ν. Since E'_ν is not real, it follows that E'_ν is not contained in $Q(\zeta_{2^{\nu+2}} + \zeta_{2^{\nu+2}}^{-1})$ and hence $E'_\nu = E_\nu$. This completes the proof of Proposition 7.5.

Corollary 7.6. The real subfields of $Q(\zeta_{2^n})$ are $Q(\zeta_{2^{\lambda+2}} + \zeta_{2^{\lambda+2}}^{-1})$, $(\lambda = 0,1,\cdots,n-2)$.

Proof. By (7.12)-(7.14), the assertion is clear.

We now return to the proof of Theorem 7.4 for the case $\ell = 2$. By Corollary 7.6, it suffices to prove $S(k) = M(k)$ for the field $k = Q(\zeta_{2^c} + \zeta_{2^c}^{-1})$, where c is an arbitrary integer $\geqq 3$. We have already shown that it is enough to construct a cyclotomic algebra B_p over k for each odd prime p with even f_p (the residue class degree of p in k/Q), such that $\mathrm{inv}_p(B_p) = \frac{1}{2}$ and $\mathrm{inv}_{p'}(B_p) = 0$, p' being any prime number $\neq p, 2$. Note that if p is a prime with $p \equiv \pm 1 \pmod{2^c}$, then p splits completely in k/Q, and so $f_p = 1$.

Suppose that p is an odd prime with $p \not\equiv \pm 1 \pmod{2^c}$. Then it is easy to see that f_p is even. Put $K = Q(\zeta_{2^c}, \zeta_p)$. Then $K = Q(\zeta_{2^c})\cdot k(\zeta_p)$ and $Q(\zeta_{2^c}) \cap k(\zeta_p) = k = Q(\zeta_{2^c} + \zeta_{2^c}^{-1})$. Let ι (resp. σ) be a generating automorphism of $K/k(\zeta_p)$ (resp.

$K/Q(\zeta_{2^c}))$. Then

$$G(K/k) = <\iota> \times <\sigma>, \quad \iota^2 = \sigma^{p-1} = 1, \quad \zeta_{2^c}^{\iota} = \zeta_{2^c}^{-1}, \quad \zeta_{2^c}^{\sigma} = \zeta_{2^c}.$$

Put

$$h_{\sigma,\iota} = \zeta_{2^c}, \quad h_{\iota,\sigma} = \zeta_{2^c}^{-1}, \quad h_{\sigma} = \zeta_{2^c}^{(p-1)2}, \quad h_{\iota} = 1.$$

It is easily verified that these h's satisfy the equations (2.16)-(2.18), and hence give rise to a cyclotomic algebra B_p over k:

$$B_p = (\beta, K/k) = \sum_{i=0}^{p-2} \sum_{j=0}^{1} Ku_{\sigma}^{i}u_{\iota}^{j},$$

$$u_{\sigma}u_{\iota} = \zeta_{2^c}u_{\iota}u_{\sigma}, \quad u_{\sigma}^{p-1} = \zeta_{2^c}^{(p-1)/2}, \quad u_{\iota}^{2} = 1.$$

For any prime number $y \neq 2$, p, we have $inv_y(B_p) = 0$, because y is unramified in K/k. To determine the p-local index of B_p we need

<u>Lemma 7.7</u>. Let p be an odd prime and f the smallest positive integer such that $p^f \equiv 1 \pmod{2^c}$, $c \geq 3$. If $p \not\equiv \pm 1 \pmod{2^c}$, then $p^f \not\equiv 1 \pmod{2^{c+1}}$.

<u>Proof</u>. Write $p \equiv (-1)^{\nu}\cdot 5^{\mu} \pmod{2^{c+1}}$, where $0 \leq \nu \leq 1$, $0 \leq \mu < 2^{c-1}$, $\mu = 2^{\lambda}\kappa$, $(2, \kappa) = 1$. Of course, $p \equiv (-1)^{\nu}\cdot 5^{\mu} \pmod{2^c}$. If $p \not\equiv \pm 1 \pmod{2^c}$, then $\lambda < c - 2$. So f is

equal to $2^{c-2-\lambda} > 1$. Consequently,

$$p^f = p^{2^{c-2-\lambda}} \equiv 5^{2^\lambda \kappa 2^{c-2-\lambda}} = 5^{\kappa 2^{c-2}} \not\equiv 1 \pmod{2^{c+1}}. \qquad \#$$

Let $\mathfrak{p}$ be a prime of k lying above p. As $p \not\equiv -1 \pmod{2^c}$, $\mathfrak{p}$ splits into two primes in $Q(\zeta_{2^c})/k$. Hence the extension $K^{\mathfrak{p}}/k_{\mathfrak{p}}$ is totally ramified and cyclic of degree $p - 1$. Put $q = N_{k/Q}(\mathfrak{p})$. Then $q = p^f$, where f is the smallest positive integer such that $p^f \equiv 1 \pmod{2^c}$. We notice that

$$u_\sigma^{p-1} = \zeta_{2^c}^{(p-1)/2} = \zeta_{q-1}^{v'}, \quad v' = \{(q-1)/2^c\}\cdot\{(p-1)/2\},$$

$$v' \equiv 0 \pmod{(p-1)/2}, \qquad v' \not\equiv 0 \pmod{p-1},$$

because by Lemma 7.7, $(q - 1)/2^c$ is an odd integer. Since $G(K^{\mathfrak{p}}/k_{\mathfrak{p}}) = \langle\sigma\rangle$, we can use the formulas (4.14), (4.15) of Appendix to Chapter 4 to conclude that the $\mathfrak{p}$-index of the cyclotomic algebra B_p is equal to $(p - 1)/(v', p - 1) = 2$. The proof of Theorem 7.4 is completed.

Using similar techniques to the ones employed in the proof of Theorem 7.4, we can determine Schur subgroups of real quadratic fields:

<u>Theorem 7.8</u> (Yamada [53]). Let $k = Q(\sqrt{m})$, where m is a square free positive integer such that there is no prime $\ell \equiv 3 \pmod 4$ dividing m. Then $S(k) = M(k)$. Namely, the elements of $S(k)$ are precisely those that have uniformly distributed

invariants with values 0 or $\frac{1}{2}$.

Proof. See [53].

We will see that if k is a real quadratic field which does not satisfy the condition in Theorem 7.5, then $S(k) = S(Q) \otimes_Q k$. The proof will be performed by using defining relations of a cyclotomic algebra (cf. Chapter 2) and the formula of index of a p-adic cyclotomic algebra, combined with the following theorem.

Theorem 7.9. Let n be a positive integer which is either odd or divisible by 4. Let k be a subfield of the cyclotomic field $Q(\zeta_n)$. Then, any cyclotomic algebra B over k is similar to a cyclotomic algebra of the form:

$$(\beta,\ Q(\zeta_n,\ \zeta_b)/k), \qquad b = 4^{\delta} p_1 p_2 \cdots p_s, \tag{7.16}$$

$$\delta = 0 \ \text{ if } \ 4 \mid n, \ \text{ and } \ \delta = 1 \ \text{ if } \ 4 \nmid n. \tag{7.17}$$

where $p_1, p_2, \cdots, p_s$ are distinct odd primes not dividing n.

To prove Theorem 7.9, we require some results.

Lemma 7. 10. Let G be a finite group and let W be a finite G-module. Let H be a normal subgroup of G. Suppose that H is cyclic. Set $N = \sum_{h \in H} h$. If the image $N(W)$ of W by N equals W^H, the subset of elements of W fixed by every element of H, then we have

$$H^2(G/H, W^H) \simeq H^2(G, W).$$

The inflation map gives this isomorphism.

Proof. Let n be any non-negative integer. As H is cyclic, the n-th cohomology group $H^n(H, W)$ depends only on n being even or odd. If n is even then $H^n(H, W) = W^H/N(W)$. Since W is finite, the Herbrand quotient of the H-module W equals 1, and so the orders of $H^n(H, W)$ and $H^{n+1}(H, W)$ are the same. Thus if $N(W) = W^H$ then $H^n(H, W) = 0$ for every non-negative integer n. (For the above arguments, see [35 , VIII, §4].) Now we have

$$0 \to H^2(G/H, W^H) \overset{\text{Inf}}{\to} H^2(G, W) \overset{\text{Res}}{\to} H^2(H, W) \qquad \text{(exact)}$$

because $H^1(H, W) = 0$ (cf. [35 , Proposition 5, p. 126]). As $H^2(H,W) = 0$, it follows that the above inflation map is an isomorphism from $H^2(G/H, W^H)$ onto $H^2(G, W)$. #

Lemma 7.11. Let p be a prime. Set $k_{p,i} = Q(\zeta_{p^i})$. If p is odd, then $\langle\zeta_{p^i}\rangle = N_{k_{p,j}/k_{p,i}}(\langle\zeta_{p^j}\rangle)$ for $1 \leqq i \leqq j$. If $p = 2$, then $\langle\zeta_{2^i}\rangle = N_{k_{2,j}/k_{2,i}}(\langle\zeta_{2^j}\rangle)$ for $2 \leqq i \leqq j$.

Proof. Wellknown.

Let E be a field and ζ a root of unity. $\omega(E(\zeta))$ denotes the group consisting of roots of unity in $E(\zeta)$. Then $\omega(E(\zeta))$ is a (multiplicative) G-submodule of $E(\zeta)^*$, where $G = G(E(\zeta)/E)$. We have a canonical homomorphism of the 2-cohomology group $H^2(G, \omega(E(\zeta)))$ into $H^2(E(\zeta)/E)$, whose image is denoted by $C(E(\zeta)/E)$. An element β of $C(E(\zeta)/E)$ is identified with a class of $Br(E)$ represented by a cyclotomic algebra of the form $(\beta, E(\zeta)/E)$, and conversely.

Proposition 7.12. Let n and m be positive integers which are either odd or divisible by 4. Let k be a subfield of the cyclotomic field $Q(\zeta_n)$. Let $\{p_1, p_2, \cdots, p_s\}$ be the set of odd primes such that $p_i | m$ and $p_i \nmid n$. If $4|m$ and $4 \nmid n$, then put $m' = 2^2 p_1 p_2 \cdots p_s$. Otherwise put $m' = p_1 p_2 \cdots p_s$. Set $L = Q(\zeta_n, \zeta_m)$ and $K = Q(\zeta_n, \zeta_{m'})$. Then by the inflation map, we have

$$C(K/k) \overset{\text{Inf}}{\simeq} C(L/k).$$

Proof. It is clear that the set of primes dividing $|\omega(K)|$ and the set of primes dividing $|\omega(L)|$ are the same, which is denoted by P. Note that $2 \in P$, because $\pm 1 \in K$. It is easily verified that the proposition is proved if we have

$$H^2(G(K/k), \omega_p(K)) \overset{\text{Inf}}{\simeq} H^2(G(L/k), \omega_p(L))$$

for each $p \in P$, where for a finite extension E of Q, $\omega_p(E)$

denotes the Sylow p-subgroup of $w(E)$. Let $n = p^a h$, $m = p^b h'$, $(p, hh') = 1$, $0 \leqq a, b$. Suppose first that $b \leqq a$. Then it is easy to see that $p \nmid [L : K]$. In this case, let $K = K_0 \subset K_1 \subset K_2 \subset \cdots \subset K_t = L$ be such that K_i/K_{i-1} are cyclic extensions $(i = 1, 2, \cdots, t)$. Since $w_p(K_0) = w_p(K_1) = \cdots = w_p(K_t)$, it follows that

$$N_{K_i/K_{i-1}}(w_p(K_i)) = (w_p(K_i))^{[K_i : K_{i-1}]} = w_p(K_i) = w_p(K_{i-1}).$$

Hence by Lemma 7.10 we have

$$H^2(G(K/k), w_p(K)) \overset{\mathrm{Inf}}{\cong} H^2(G(K_1/k), w_p(K_1))$$
$$\overset{\mathrm{Inf}}{\cong} \cdots \overset{\mathrm{Inf}}{\cong} H^2(G(L/k), w_p(L)).$$

Suppose next that $b > a$. If $p \neq 2$, put $c = \mathrm{Max}\{a, 1\}$. If $p = 2$, put $c = \mathrm{Max}\{a, 2\}$. Set $E = Q(\zeta_n, \zeta_{m/p^{b-c}})$. Then $p \nmid [E : K]$, $w_p(E) = w_p(K) = \langle\zeta_{p^c}\rangle$, $L = E(\zeta_{p^b})$ and $G(L/E)$ is canonically isomorphic to $G(Q(\zeta_{p^b})/Q(\zeta_{p^c}))$. It follows from Lemmas 7.10, 7.11 and the same argument as in the case $b \leqq a$ that

$$H^2(G(K/k), w_p(K)) \overset{\mathrm{Inf}}{\cong} H^2(G(E/k), w_p(E)) \overset{\mathrm{Inf}}{\cong} H^2(G(L/k), w_p(L)). \quad \#$$

<u>Corollary 7.13</u>. Notation being the same as in Proposition 7.12, Let $B = (\beta, k(\zeta_m)/k)$ be a cyclotomic algebra over k.

Then B is similar to a cyclotomic algebra of the form $(\alpha, Q(\zeta_n, \zeta_{m'})/k)$.

<u>Proof</u>. Note that $B \sim (\mathrm{Inf}(\beta), L/k)$, where $\mathrm{Inf} = \mathrm{Inf}_{k(\zeta_m)\to L}$. Since $\mathrm{Inf}(\beta) \in C(L/k)$, Proposition 7.12 yields that there exists $\alpha \in C(K/k)$ such that $\mathrm{Inf}_{K\to L}(\alpha) = \mathrm{Inf}(\beta)$, and hence $B \sim (\mathrm{Inf}(\beta), L/k) \sim (\alpha, K/k)$. #

Now we are ready to prove Theorem 7.9.

<u>Proof of Theorem 7.9</u>. Let $B = (\beta, k(\zeta_m)/k)$ be a cyclotomic algebra over k. By Corollary 7.13, we may assume that B is of the form:

$$B = (\alpha, Q(\zeta_n, \zeta_{m'})/k), \quad m' = 4^{\delta} p_1 p_2 \cdots p_s,$$

where $p_1, p_2, \cdots, p_s$ are distinct odd primes not dividing n, $\delta = 0$ or 1, and $\delta = 0$ if $4 | n$. In the case $4 \nmid n$, $\delta = 0$, put $m'' = 4 p_1 p_2 \cdots p_s$. Then $B \sim (\mathrm{Inf}(\alpha), Q(\zeta_n, \zeta_{m''})/k)$, where Inf denotes the inflation map from $H^2(Q(\zeta_n, \zeta_{m'})/k)$ into $H^2(Q(\zeta_n, \zeta_{m''})/k)$. The assertion of Theorem 7.9 follows immediately.

The determination of the Schur subgroup of any arbitrary real quadratic field will be completed by the next theorem.

<u>Theorem 7.14</u> (Yamada [55]). Let $k = Q(\sqrt{m})$, where m is a square free positive integer divisible by a prime $\ell \equiv 3 \pmod 4$.

Then $S(k) = S(Q) \otimes_Q k$. In other words, $S(k)$ consists of the classes of $Br(k)$ which have uniformly distributed invariants with values 0 or $\frac{1}{2}$ such that the p-local invariant is 0 whenever p does not split in k.

Proof. We write

$$m = \ell_1 \ell_2 \cdots \ell_h, \quad \text{if } 2 \nmid m, \tag{7.18}$$

$$m = 2\ell_1 \ell_2 \cdots \ell_h, \quad \text{if } 2 \mid m, \tag{7.19}$$

where $\ell_1, \ell_2, \cdots, \ell_h$ are distinct primes and

$$h = n + n', \quad 1 \leqq n, \quad 0 \leqq n', \tag{7.20}$$

$$\ell_i \equiv 3 \pmod 4 \quad \text{for } 1 \leqq i \leqq n, \tag{7.21}$$

$$\ell_j \equiv 1 \pmod 4 \quad \text{for } n < j \leqq h. \tag{7.22}$$

Put $d = 4m$ (resp. $d = m$), if $m \equiv 2, 3 \pmod 4$ (resp. $m \equiv 1 \pmod 4$). Then d is the discriminant of k. It is wellknown that the field $Q(\zeta_d)$ of d-th roots of unity is the minimal cyclotomic field containing k ([29, §27, 4, p.513]). Since the Schur subgroup $S(k)$ consists of those classes of $Br(k)$ that contain a cyclotomic algebra over k, the theorem will be established if we shall prove that a cyclotomic algebra over k has p-local invariant 0 (or p-local index 1) whenever p does not split in k.

Now we apply Theorem 7.9 to the real quadratic field $k = Q(\sqrt{m}) \subset Q(\zeta_d)$ and a cyclotomic algrbra B over k. This yields that any arbitrary cyclotomic algebra B over k may be assumed to be of the form:

$$B = (\beta, L/k) = \sum_{\sigma\varepsilon G} Lu_\sigma, \quad (u_1 = 1), \quad G = G(L/k), \tag{7.23}$$

$$u_\sigma u_\tau = \beta(\sigma, \tau)u_{\sigma\tau}, \quad u_\sigma x = x^\sigma u_\sigma \quad (x \varepsilon L), \tag{7.24}$$

$$L = Q(\zeta_d, \zeta_b), \quad b = 4^\delta p_1 p_2 \cdots p_t, \tag{7.25}$$

where $\delta = 0$ (resp. $\delta = 1$) if $4|d$ (resp. $4\nmid d$), and where $p_1, p_2, \cdots, p_t$ are distinct odd primes not dividing d. Let $W_2(L)$ denote the group of roots of unity contained in L whose orders are powers of 2. We have

$$W_2(L) = \langle\zeta_4\rangle \text{ (resp. } \langle\zeta_8\rangle), \quad \text{if } 2\nmid m \text{ (resp. if } 2|m). \tag{7.26}$$

Because the index of B divides 2, we may assume that

$$\beta(\sigma, \tau) \,\varepsilon\, W_2(L) \quad \text{for all } \sigma, \tau \,\varepsilon\, G(L/k). \tag{7.27}$$

We will show that $inv_p(B) = 0$ whenever p does not split in k. The rational infinite p_∞ splits in k. Let P denote the set of prime numbers that are ramified in L/Q. Then

$$P = \{2, \ell_1, \ell_2, \cdots, \ell_h, p_1\ p_2, \cdots, p_t\}. \tag{7.28}$$

If a prime number p does not belong to P, then $\mathrm{inv}_p(B) = 0$, because p is unramified in L/k and the values of the factor set β are roots of unity. Since the ramification index of ℓ_i in k/Q is 2, it follows from Theorem 4.3 that the ℓ_i-local index of B divides $(\ell_i - 1)/2$, which is odd for $i = 1, 2, \cdots, n$ ($\ell_i \equiv 3 \pmod 4$ for $1 \leq i \leq n$). On the other hand, the index of B divides 2. Thus we conclude that

$$\mathrm{inv}_{\ell_i}(B) = 0 \quad \text{for } i = 1, 2, \cdots, n. \qquad (7.29)$$

We will prove

$$\mathrm{inv}_{\ell_j}(B) = 0, \qquad (n < j \leq h), \qquad (7.30)$$

$$\mathrm{inv}_{p_\nu}(B) = 0, \text{ if } p_\nu \text{ is inertial in } k, (1 \leq \nu \leq t). \quad (7.31)$$

(If p_ν is not inertial, then p_ν splits into two primes in k.) By virtue of Hasse's sum theorem it follows that $\mathrm{inv}_2(B) = 0$, if 2 is either inertial or ramified in k. (We can also prove this directly, using the formula of invariant of a 2-adic cyclotomic algebra given in Theorem 5.6.) Therefore, if we will establish (7.30) and (7.31), then $\mathrm{inv}_p(B) = 0$ whenever p does not split in k, and thus the proof of the theorem will be completed.

In order to prove (7.30), (7.31) we will extensively use the results in Appendix to Chapter 4, where a simple process for determining local indices of a cyclotomic algebra is given. (In particular, we will make use of Proposition 4.9.)

Case I. $k = Q(\sqrt{m})$, $2\nmid m$, $2\nmid n$.

Suppose that $2\nmid m$ and $2\nmid n$ in (7.18)-(7.22). Then $m \equiv 3 \pmod 4$ and $d = 4m$. We have

$$G(Q(\zeta_d)/Q)) = \langle\iota\rangle \times \langle\psi_1\rangle \times \langle\psi_2\rangle \times \cdots \times \langle\psi_h\rangle, \tag{7.32}$$

$$\zeta_4^{\iota} = \zeta_4^{-1},\quad \zeta_{d/4}^{\iota} = \zeta_{d/4},\quad \zeta_{\ell_i}^{\psi_i} = \zeta_{\ell_i}^{r_i},\quad \zeta_{d/\ell_i}^{\psi_i} = \zeta_{d/\ell_i}, \tag{7.33}$$

where r_i is a primitive root modulo ℓ_i $(i = 1, 2, \cdots, h)$. Put $\ell_i^* = -\ell_i$ for $1 \leq i \leq n$ $(\ell_i \equiv 3 \pmod 4)$, and $\ell_i^* = \ell_i$ for $n < i \leq h$ $(\ell_i \equiv 1 \pmod 4)$. Then $Q(\sqrt{\ell_i^*})$ is the unique quadratic extension of Q in $Q(\zeta_{\ell_i})$, $(i = 1, 2, \cdots, h)$, (cf. [29, III, §27, 4]), and so $\psi_i(\sqrt{\ell_i^*}) = -\sqrt{\ell_i^*}$ $(1 \leq i \leq h)$. Hence

$$\iota^{\nu}\psi_1^{\nu_1}\cdots\psi_h^{\nu_h}(\sqrt{m}) = (-1)^{\nu+\nu_1+\cdots+\nu_h}\cdot\sqrt{m},$$

because $\sqrt{m} = \sqrt{-1}\sqrt{-\ell_1}\cdots\sqrt{-\ell_n}\sqrt{\ell_{n+1}}\cdots\sqrt{\ell_h}$. This implies that an automorphism $\iota^{\nu}\psi_1^{\nu_1}\cdots\psi_h^{\nu_h}$ of $Q(\zeta_d)/Q$ belongs to $G(Q(\zeta_d)/k)$ if and only if $\nu + \nu_1 + \cdots + \nu_h \equiv 0 \pmod 2$. We conclude from this that

$$G(Q(\zeta_d)/k) = \langle\phi_1\rangle \times \langle\phi_2\rangle \times \cdots \times \langle\phi_h\rangle,\quad \phi_i = \iota\psi_i, \tag{7.34}$$

$$|\langle\phi_i\rangle| = \ell_i - 1, \quad \phi_i(\zeta_4) = \zeta_4^{-1}, \quad (i = 1,2,\cdots,h). \tag{7.35}$$

Let

$$B = (\beta,\ L/k) = \sum_{\sigma\varepsilon G} Lu_\sigma, \quad (u_1 = 1),$$

$$L = Q(\zeta_d,\ \zeta_b), \quad b = p_1p_2\cdots p_t, \quad w_2(L) = \langle\zeta_4\rangle$$

be a cyclotomic algebra over k defined by (7.23)-(7.25), (7.27). We have

$$G(L/k) = \prod_{i=1}^{h} \langle\phi_i\rangle \times \prod_{\nu=1}^{t} \langle\xi_\nu\rangle, \tag{7.36}$$

where ϕ_i is given by (7.34), (7.35) together with $\phi_i(\zeta_b) = \zeta_b$ $(1 \leq i \leq h)$, and

$$\xi_\nu(\zeta_{p_\nu}) = \zeta_{p_\nu}^{s_\nu}, \quad \xi_\nu(\zeta_{db/p_\nu}) = \zeta_{db/p_\nu}, \quad |\langle\xi_\nu\rangle| = p_\nu-1, \tag{7.37}$$

s_ν being a primitive root modulo p_ν $(1 \leq \nu \leq t)$. Note that $\xi_\nu(\zeta_4) = \zeta_4$. By the results of Chapter 2, we conclude that

$$B = (\beta,\ L/k) = (\beta',\ L/k) = \sum Lu_{\phi_1}^{\nu_1}\cdots u_{\phi_h}^{\nu_h}u_{\xi_1}^{\mu_1}\cdots u_{\xi_t}^{\mu_t}, \tag{7.38}$$

$$\beta'(\sigma,\ \sigma') \in \langle\zeta_4\rangle \quad \text{for all} \quad \sigma,\ \sigma' \in G(L/k), \tag{7.39}$$

where the sum is taken over $0 \leq \nu_i < \ell_i - 1$ $(i = 1,2,\cdots,h)$ and $0 \leq \mu_j < p_j - 1$ $(j = 1,2,\cdots,t)$. (The definition of the factor set β' is as in Chapter 2.) We have

$$u_{\phi_i}u_{\phi_j} = \zeta_4^{x_{ij}}u_{\phi_j}u_{\phi_i}, \quad u_{\phi_i}u_{\xi_\nu} = \zeta_4^{y_{i\nu}}u_{\xi_\nu}u_{\phi_i}, \tag{7.40}$$

$$u_{\xi_\nu}u_{\xi_\mu} = \zeta_4^{z_{\nu\mu}}u_{\xi_\mu}u_{\xi_\nu}, \quad u_{\phi_i}^{\ell_i-1} = \zeta_4^{b_i}, \quad u_{\xi_\nu}^{p_\nu-1} = \zeta_4^{c_\nu}, \tag{7.41}$$

where x_{ij}, $y_{i\nu}$, $z_{\nu\mu}$, b_i, c_ν are some integers $(1 \leqq i, j \leqq h;$ $1 \leqq \nu, \mu \leqq t)$. (The cyclotomic algebra B has other defining relations which are unnecessary for computing local indices of B.)

Now we will prove $\mathrm{inv}_{\ell_j}(B) = 0$ for each $j = n + 1, \cdots, h$. The inertia group of ℓ_j in L/k is $\langle\phi_j^2\rangle$ $(= \langle\psi_j^2\rangle)$. It follows from (7.40), (7.35) that for each $i = 1, 2, \cdots, h$,

$$u_{\phi_j}^2 u_{\phi_i} = u_{\phi_j}\zeta_4^{x_{ji}}u_{\phi_i}u_{\phi_j} = \zeta_4^{-x_{ji}}u_{\phi_j}u_{\phi_i}u_{\phi_j} = u_{\phi_i}u_{\phi_j}^2,$$

$(1 \leqq j \leqq h)$. In the same way, we have $u_{\phi_j}^2 u_{\xi_\nu} = u_{\xi_\nu}u_{\phi_j}^2$, $(1 \leqq \nu \leqq t)$. Let $\eta = \phi_1^{a_1}\cdots\phi_h^{a_h}\xi_1^{a'_1}\cdots\xi_t^{a'_t}$ be a Frobenius automorphism of ℓ_j in L/k, where a_i, a'_ν are some integers such that $0 \leqq a_i < \ell_i - 1$ $(1 \leqq i \leqq h)$, $0 \leqq a'_\nu < p_\nu - 1$ $(1 \leqq \nu \leqq t)$. Put

$$\omega = \phi_j^2, \quad v_\omega = u_{\phi_j}^2, \quad v_\eta = u_{\phi_1}^{a_1}\cdots u_{\phi_h}^{a_h}u_{\xi_1}^{a'_1}\cdots u_{\xi_t}^{a'_t}.$$

Since $u_{\phi_j}^2$ commutes with u_{ϕ_i} and u_{ξ_ν}, we have $v_\omega v_\eta = v_\eta v_\omega$.

Because $\phi_j(\zeta_4) = \zeta_4^{-1}$, it follows from the relation (2.16) that

$$u_{\phi_j}^{\ell_j - 1} = (-1)^{b'_j} = \zeta_{\ell_j - 1}^{v'}, \quad v' = b'_j(\ell_j - 1)/2$$

for some integer b'_j. We notice that for our case, the conditions of Proposition 4.9 are satisfied. Hence, by the formula (4.27) we conclude the ℓ_j-local index of B is equal to

$$\frac{(\ell_j - 1)/2}{(v', (\ell_j - 1)/2)} = 1.$$

(The ramification index and the residue classdegree of ℓ_j in k/Q are, respectively, 2 and 1.)

Next we will prove that if p_ν $(1 \leq \nu \leq t)$ is inertal in k, then $\mathrm{inv}_{p_\nu}(B) = 0$. By (7.40), (7.35), we have

$$u_{\xi_\nu} u_{\phi_i}^2 = \zeta_4^{-y_{i\nu}} u_{\phi_i} u_{\xi_\nu} u_{\phi_i} = \zeta_4^{-y_{i\nu}} u_{\phi_i} \zeta_4^{-y_{i\nu}} u_{\phi_i} u_{\xi_\nu} = u_{\phi_i}^2 u_{\xi_\nu},$$

$(1 \leq i \leq h)$. The relation (2.19), when applied to the automorphisms ξ_ν, ξ_μ, ϕ_1, yields that

$$1 = (\zeta_4^{z_{\nu\mu}})^{\phi_1 - 1} (\zeta_4^{-y_{1\mu}})^{\xi_\nu - 1} (\zeta_4^{y_{1\nu}})^{\xi_\mu - 1} = \zeta_4^{-2z_{\nu\mu}},$$

$(1 \leq \nu, \mu \leq t;\ \nu \neq \mu)$. Hence $2 | z_{\nu\mu}$. It follows that $u_{\xi_\nu} u_{\xi_\mu} = \pm u_{\xi_\mu} u_{\xi_\nu}$, and so $u_{\xi_\nu} u_{\xi_\mu}^2 = u_{\xi_\mu}^2 u_{\xi_\nu}$. Let

$$p_\nu^2 \equiv r_i^{a_i} \pmod{\ell_i}, \quad 0 \leq a_i < \ell_i - 1, \quad (1 \leq i \leq h), \qquad (7.42)$$

$$p_\nu^2 \equiv s_\mu^{a'_\mu} \pmod{p_\mu}, \quad 0 \leq a'_\mu < p_\mu - 1, \quad (1 \leq \mu \leq t;\ \mu \neq \nu). \quad (7.43)$$

($p_\nu^2 \equiv 1 \pmod 4$). The integers a_i, a'_μ are uniquely determined.)
Then $2|a_i$, $2|a'_\mu$. Put

$$\eta = \prod_{i=1}^{h} \phi_i^{a_i} \prod_{\substack{\mu=1 \\ \mu\neq\nu}}^{t} \xi_\mu^{a'_\mu},$$

$$v_\eta = u_{\phi_1}^{a_1} \cdots u_{\phi_h}^{a_h} u_{\xi_1}^{a'_1} \cdots u_{\xi_{\nu-1}}^{a'_{\nu-1}} u_{\xi_{\nu+1}}^{a'_{\nu+1}} \cdots u_{\xi_t}^{a'_t}.$$

It is easy to see that η is a Frobenius automorphism of p_ν in L/k. The inertia group of p_ν in L/k is $\langle\xi_\nu\rangle$. Since u_{ξ_ν} commutes with $u_{\phi_i}^2$ $(1 \leq i \leq h)$ and $u_{\xi_\mu}^2$ $(1 \leq \mu \leq t;\ \mu \neq \nu)$, and $2|a_i$, $2|a'_\mu$, it follows that $u_{\xi_\nu} v_\eta = v_\eta u_{\xi_\nu}$, i.e., $\beta'(\xi_\nu, \eta)/\beta'(\eta, \xi_\nu) = 1$. Referring to the relation (2.18) we have

$$\zeta_4^{-2c_\nu} = (\zeta_4^{c_\nu})^{\phi_1 - 1} = (\zeta_4^{y_{1\nu}})^{1+\xi_\nu+\cdots+\xi_\nu^{p_\nu-2}} = \zeta_4^{y_{1\nu}(p_\nu-1)},$$

whence $c_\nu = -y_{1\nu}(p_\nu - 1)/2 + 2z$ for some integer z. Therefore, we have

$$u_{\xi_\nu}^{p_\nu-1} = \zeta_4^{-y_{1\nu}(p_\nu-1)/2+2z} = \zeta_{q-1}^{v'}, \quad (\zeta_4 = \zeta_{q-1}^{(q-1)/4}),$$

$$v' = -y_{1\nu}(p_\nu-1)(q-1)/8 + z(p_\nu-1)(p_\nu+1)/2 \equiv 0 \pmod{p_\nu - 1},$$

where $q = p_\nu^2 = N_{k/Q}(\mathfrak{p}_\nu)$, $\mathfrak{p}_\nu$ is the prime of k dividing p_ν, and $q - 1 = p_\nu^2 - 1 \equiv 0 \pmod 8$, $p_\nu + 1 \equiv 0 \pmod 2$. Recall that p_ν is unramified in k/Q. We can now apply Proposition 4.9 to our case. By the formula (4.27), we conclude the p_ν-local index of B is equal to $(p_\nu - 1)/(v', p_\nu - 1) = 1$, because $v' \equiv 0 \pmod{p_\nu - 1}$. This completes the proof of the theorem for the case $2 \nmid m$, $2 \nmid n$.

<u>Case II</u>. $k = Q(\sqrt{m})$, $2 \nmid m$, $2 | n$.

Suppose that $2 \nmid m$ and $2 | n$ in (7.18)-(7.22). Then $m \equiv 1 \pmod 4$, and so $d = m$. We have

$$G(Q(\zeta_d)/Q) = \prod_{i=1}^{h} \langle \psi_i \rangle, \quad \zeta_{\ell_i}^{\psi_i} = \zeta_{\ell_i}^{r_i}, \quad \zeta_{m/\ell_i}^{\psi_i} = \zeta_{m/\ell_i}, \qquad (7.44)$$

r_i being a primitive root modulo ℓ_i. By the same argument as in Case I, we conclude that

$$\psi_1^{\nu_1} \cdots \psi_h^{\nu_h}(\sqrt{m}) = (-1)^{\nu_1 + \cdots + \nu_h}(\sqrt{m}),$$

because $\sqrt{m} = \sqrt{-\ell_1} \cdots \sqrt{-\ell_n}\sqrt{\ell_{n+1}} \cdots \sqrt{\ell_h}$. Hence $\psi_1^{\nu_1} \cdots \psi_h^{\nu_h}$ belongs to $G(Q(\zeta_d)/k)$ if and only if $\nu_1 + \cdots + \nu_h \equiv 0 \pmod 2$. Put $g = (\ell_1 - 1)/2$. Then $2 \nmid g$. Set

$$\phi_1 = \psi_1^2, \quad \phi_i = \psi_1^g \psi_i \ (i = 2,3,\cdots,h), \quad (\phi_i^2 = \psi_i^2). \qquad (7.45)$$

It is easy to see that

$$G(Q(\zeta_d)/k) = \langle \phi_1 \rangle \times \langle \phi_2 \rangle \times \cdots \times \langle \phi_h \rangle,$$

$$|\langle\phi_1\rangle| = (\ell_1 - 1)/2, \quad |\langle\phi_j\rangle| = \ell_j - 1 \ (2 \leq j \leq h).$$

Let $B = (\beta, L/k)$, $L = Q(\zeta_d, \zeta_b)$, $b = 4p_1p_2\cdots p_t$ be a cyclotomic algebra over k defined by (7.23)-(7.25), (7.27). We have

$$G(L/k) = \prod_{i=1}^{h} \langle\phi_i\rangle \times \langle\iota\rangle \times \prod_{\nu=1}^{t} \langle\xi_\nu\rangle, \tag{7.46}$$

where ϕ_i is defined by (7.44), (7.45) together with $\phi_i(\zeta_b) = \zeta_b$ $(1 \leq i \leq h)$, and ξ_ν is defined by (7.37), $(1 \leq \nu \leq t)$, and $\iota(\zeta_4) = \zeta_4^{-1}$, $\iota(\zeta_{db/4}) = \zeta_{db/4}$. Then,

$$B = (\beta, L/k) = (\beta', L/k) = \Sigma\, L u_{\phi_1}^{\nu_1}\cdots u_{\phi_h}^{\nu_h} u_\iota^{\mu} u_{\xi_1}^{\mu_1}\cdots u_{\xi_t}^{\mu_t}, \tag{7.47}$$

where the sum is taken over $0 \leq \nu_1 < (\ell_1 - 1)/2$, $0 \leq \nu_i < \ell_i - 1$ $(i = 2,\cdots,h)$, $0 \leq \mu \leq 1$, and $0 \leq \mu_j < p_j - 1$ $(j = 1,\cdots,t)$. We have

$$u_{\phi_i}u_{\phi_j} = \zeta_4^{x_{ij}}u_{\phi_j}u_{\phi_i}, \quad u_{\phi_i}u_\iota = \zeta_4^{x_i}u_\iota u_{\phi_i}, \quad u_{\phi_i}u_{\xi_\nu} = \zeta_4^{y_{i\nu}}u_{\xi_\nu}u_{\phi_i}, \tag{7.48}$$

$$u_\iota u_{\xi_\nu} = \zeta_4^{y_\nu}u_{\xi_\nu}u_\iota, \quad u_{\xi_\nu}u_{\xi_\mu} = \zeta_4^{z_{\nu\mu}}u_{\xi_\mu}u_{\xi_\nu}, \tag{7.49}$$

$$u_{\phi_1}^{(\ell_1-1)/2} = \zeta_4^{b_1}, \quad u_{\phi_i}^{\ell_i-1} = \zeta_4^{b_i} \ (i \neq 1), \quad u_{\xi_\nu}^{p_\nu-1} = \zeta_4^{c_\nu}, \tag{7.50}$$

where $1 \leq i,\ j \leq h$, $1 \leq \nu,\ \mu \leq t$. Note that $\phi_i(\zeta_4) = \xi_\nu(\zeta_4) = \zeta_4$. Referring to the relation (2.19), we have

$$1 = (\zeta_4^{x_{ij}})^{\iota-1}(\zeta_4^{x_j})^{\phi_i-1}(\zeta_4^{-x_i})^{\phi_j-1} = \zeta_4^{-2x_{ij}} = (-1)^{x_{ij}},$$

$$1 = (\zeta_4^{y_{i\nu}})^{\iota-1}(\zeta_4^{-y_\nu})^{\phi_i-1}(\zeta_4^{-x_i})^{\xi_\nu-1} = \zeta_4^{-2y_{i\nu}} = (-1)^{y_{i\nu}},$$

$$1 = (\zeta_4^{z_{\nu\mu}})^{\iota-1}(\zeta_4^{-y_\mu})^{\xi_\nu-1}(\zeta_4^{y_\nu})^{\xi_\mu-1} = \zeta_4^{-2z_{\nu\mu}} = (-1)^{z_{\nu\mu}}.$$

It follows that x_{ij}, $y_{i\nu}$, $z_{\nu\mu}$ are even numbers, and so

$$u_{\phi_i}u_{\phi_j} = \pm u_{\phi_j}u_{\phi_i}, \quad u_{\phi_i}u_{\xi_\nu} = \pm u_{\xi_\nu}u_{\phi_i}, \quad u_{\xi_\nu}u_{\xi_\mu} = \pm u_{\xi_\mu}u_{\xi_\nu}. \qquad (7.51)$$

We will prove $\mathrm{inv}_{\ell_j}(B) = 0$ for $j = n + 1, \cdots, h$. The inertia group of ℓ_j in L/k is $\langle\phi_j^2\rangle$ $(= \langle\psi_j^2\rangle)$. Since $\ell_j \equiv 1 \pmod 4$, it follows that any Frobenius automorphism η of ℓ_j in L/k is of the form: $\eta = \phi_1^{a_1}\cdots\phi_h^{a_h}\xi_1^{a'_1}\cdots\xi_t^{a'_t}$. Since by (7.51) $u_{\phi_j}^2$ commutes with u_{ϕ_i} $(1 \leq i \leq h)$ and u_{ξ_ν} $(1 \leq \nu \leq t)$, we conclude $\beta'(\phi_j^2, \eta)/\beta'(\eta, \phi_j^2) = 1$. Referring to the relation (2.18), we have

$$(-1)^{b_j} = \zeta_4^{-2b_j} = (\zeta_4^{b_j})^{\iota-1} = (\zeta_4^{-x_j})^{1+\phi_j+\cdots+\phi_j^{\ell_j-2}}$$

$$= \zeta_4^{-x_j(\ell_j-1)} = 1,$$

$(\ell_j \equiv 1 \pmod 4)$. Hence $2|b_j$, and consequently, $u_{\phi_j}^{\ell_j-1} = \pm 1$. By using Proposition 4.9 as in Case I, we conclude that the ℓ_j-local index of B is equal to 1, as requested.

Next we will prove that if p_ν is inertial in k/Q, then $\mathrm{inv}_{p_\nu}(B) = 0$. The inertia group of p_ν in L/k is $\langle\xi_\nu\rangle$. By (7.51), $u_{\xi_\nu}u_{\phi_1} = (-1)^\kappa u_{\phi_1}u_{\xi_\nu}$, where κ is some integer. Referring to the relation (2.18), we have

$$1 = (\zeta_4^{b_1})^{\xi_\nu - 1} = ((-1)^\kappa)^{1+\phi_1+\cdots+\phi_1^{(\ell_1-1)/2-1}}$$

$$= (-1)^{\kappa(\ell_1-1)/2} = (-1)^\kappa,$$

because $(\ell_1 - 1)/2$ is odd. Consequently, $2|\kappa$, and so u_{ξ_ν} commutes with u_{ϕ_1}. By (7.51), u_{ξ_ν} also commutes with $u_{\phi_i}^2$ $(2 \leqq i \leqq h)$ and with $u_{\xi_\mu}^2$ $(1 \leqq \mu \leqq t)$. Let a_i $(1 \leqq i \leqq h)$ and a'_μ $(1 \leqq \mu \leqq t;\ \mu \neq \nu)$ be the integers defined by (7.42), (7.43). Then $2|a_i$, $2|a'_\mu$. It is easy to see that

$$\eta = \phi_1^{a_1/2} \prod_{i=2}^{h} \phi_i^{a_i} \prod_{\substack{\mu=1 \\ \mu\neq\nu}}^{t} \xi_\mu^{a'_\mu}$$

is a Frobenius automorphism of p_ν in L/k $(p_\nu^2 \equiv 1 \pmod 4)$, and that $\beta'(\xi_\nu, \eta)/\beta'(\eta, \xi_\nu) = 1$. Referring to the relation (2.18), we have

$$\zeta_4^{-2c_\nu} = (\zeta_4^{c_\nu})^{\iota-1} = (\zeta_4^{y_\nu})^{1+\xi_\nu+\cdots+\xi_\nu^{p_\nu-2}} = \zeta_4^{y_\nu(p_\nu-1)},$$

whence

$$u_{\xi_\nu}^{p_\nu-1} = \zeta_4^{c_\nu} = \zeta_4^{-y_\nu(p_\nu-1)/2+2z}$$

for some integer z. Then by the same argument as in Case I, we conclude that the p_ν-local index of B is equal to 1. This completes the proof of the theorem for the case $2\nmid m$, $2|n$.

<u>Case III</u>. $k = Q(\sqrt{m})$, $2|m$.

Suppose that $2|m$. Then $d = 4m \equiv 0 \pmod 8$. We have

$$G(Q(\zeta_d)/Q) = \langle\iota\rangle \times \langle\theta\rangle \times \prod_{i=1}^{h} \langle\psi_i\rangle, \tag{7.52}$$

$$\zeta_8^\iota = \zeta_8^{-1}, \quad \zeta_8^\theta = \zeta_8^5, \quad \zeta_{d/8}^\iota = \zeta_{d/8}^\theta = \zeta_{d/8}, \tag{7.53}$$

$$\zeta_{\ell_i}^{\psi_i} = \zeta_{\ell_i}^{r_i}, \quad \zeta_{d/\ell_i}^{\psi_i} = \zeta_{d/\ell_i}, \quad (1 \leqq i \leqq h), \tag{7.54}$$

where r_i is a primitive root modulo ℓ_i. We see that

$$\sqrt{m} = \sqrt{2}\sqrt{-\ell_1}\cdots\sqrt{-\ell_n}\sqrt{\ell_{n+1}}\cdots\sqrt{\ell_h}, \quad \text{if } 2|n, \tag{7.55}$$

$$\sqrt{m} = \sqrt{-2}\sqrt{-\ell_1}\cdots\sqrt{-\ell_n}\sqrt{\ell_{n+1}}\cdots\sqrt{\ell_h}, \quad \text{if } 2\nmid n. \tag{7.56}$$

Note that $\iota(\sqrt{2}) = \sqrt{2}$, $\theta(\sqrt{2}) = -\sqrt{2}$, $\iota\theta(\sqrt{-2}) = \sqrt{-2}$, $\theta(\sqrt{-2}) = -\sqrt{-2}$. If $2|n$, then

$$\iota^\nu\theta^{\nu'}\phi_1^{\nu_1}\cdots\phi_h^{\nu_h}(\sqrt{m}) = (-1)^{\nu'+\nu_1+\cdots+\nu_h}\cdot\sqrt{m}.$$

From this it follows easily that

$$G(Q(\zeta_d)/k) = \langle\iota\rangle \times \langle\theta\psi_1\rangle \times \cdots \times \langle\theta\psi_h\rangle.$$

Similarly, if $2 \nmid n$, then

$$G(Q(\zeta_d)/k) = \langle \iota\theta \rangle \times \langle \theta\psi_1 \rangle \times \cdots \times \langle \theta\psi_h \rangle.$$

Setting

$$\rho = \iota \ \text{ for } \ 2 | n, \ \text{ and } \ \rho = \iota\theta \ \text{ for } \ 2 \nmid n, \tag{7.57}$$

$$\theta\psi_i = \phi_i, \quad (i = 1, 2, \cdots, h), \tag{7.58}$$

we have

$$G(Q(\zeta_d)/k) = \langle \rho \rangle \times \langle \phi_1 \rangle \times \langle \phi_2 \rangle \times \cdots \times \langle \phi_h \rangle, \tag{7.59}$$

$$\rho(\zeta_8) = \begin{cases} \zeta_8^{-1}, & \text{for } 2|n, \\ \zeta_8^3, & \text{for } 2\nmid n, \end{cases} \qquad \phi_i(\zeta_8) = \zeta_8^5 \ (1 \leqq i \leqq h). \tag{7.60}$$

Let $B = (\beta, L/k) = \sum_\sigma Lu_\sigma$, $L = Q(\zeta_d, \zeta_b)$, $b = p_1 p_2 \cdots p_t$ be a cyclotomic algebra over k defined by (7.23)-(7.25), (7.27). The Galois group of the extension L/k is:

$$G(L/k) = \langle \rho \rangle \times \prod_{i=1}^{h} \langle \phi_i \rangle \times \prod_{\nu=1}^{t} \langle \xi_\nu \rangle,$$

where ρ and ϕ_i are defined as above, together with $\rho(\zeta_b) = \phi_i(\zeta_b) = \zeta_b$, and ξ_ν is defined by (7.37), $(1 \leqq \nu \leqq t)$. We have

$$B = (\beta, L/k) = (\beta', L/k) = \sum Lu_\rho^{\nu} u_{\phi_1}^{\nu_1} \cdots u_{\phi_h}^{\nu_h} u_{\xi_1}^{\mu_1} \cdots u_{\xi_t}^{\mu_t}, \tag{7.61}$$

where the sum is taken over $\nu = 0, 1$, $0 \leqq \nu_i < \ell_i - 1$ ($i = 1, \cdots, h$), $0 \leqq \mu_j < p_j - 1$ ($j = 1, \cdots, t$). We have

$$u_\rho u_{\phi_i} = \zeta_8^{x_i} u_{\phi_i} u_\rho, \quad u_\rho u_{\xi_\nu} = \zeta_8^{y_\nu} u_{\xi_\nu} u_\rho, \quad u_{\phi_i} u_{\phi_j} = \zeta_8^{x_{ij}} u_{\phi_j} u_{\phi_i}, \tag{7.62}$$

$$u_{\phi_i} u_{\xi_\nu} = \zeta_8^{x'_{i\nu}} u_{\xi_\nu} u_{\phi_i}, \quad u_{\xi_\nu} u_{\xi_\mu} = \zeta_8^{y_{\nu\mu}} u_{\xi_\mu} u_{\xi_\nu}, \tag{7.63}$$

$$u_\rho^2 = (-1)^z, \quad u_{\phi_i}^{\ell_i - 1} = \zeta_4^{z_i}, \quad u_{\xi_\nu}^{p_\nu - 1} = \zeta_8^{z'_\nu} \tag{7.64}$$

where $1 \leqq i, j \leqq h$, $1 \leqq \nu, \mu \leqq t$. (Because $\rho(\zeta_4) = \zeta_4^{-1} \neq \zeta_4$, $\phi_i(\zeta_8) = \zeta_8^5 \neq \zeta_8$, $\phi_i(\zeta_4) = \zeta_4$, it follows that $u_\rho^2 = \pm 1$, $u_{\phi_i}^{\ell_i - 1} = \zeta_4^{z_i}$ for some integer z_i.) Referring to the relation (2.18), we have

$$(-1)^{z_1} = \zeta_4^{\pm 2 z_1} = (\zeta_4^{z_1})^{\rho - 1} = (\zeta_8^{x_1})^{1 + \phi_1 + \cdots + \phi_1^{\ell_1 - 2}} = \zeta_8^{x_1 S}, \tag{7.65}$$

$$1 = (\zeta_4^{z_1})^{\phi_i - 1} = (\zeta_8^{-x_{1i}})^{1 + \phi_1 + \cdots + \phi_1^{\ell_1 - 2}} = \zeta_8^{-x_{1i} S}, \quad (i \neq 1) \tag{7.66}$$

$$1 = (\zeta_4^{z_1})^{\xi_\nu - 1} = (\zeta_8^{-x'_{1\nu}})^{1 + \phi_1 + \cdots + \phi_1^{\ell_1 - 2}} = \zeta_8^{-x'_{1\nu} S}, \tag{7.67}$$

$$S = 1 + 5 + \cdots + 5^{\ell_1 - 2} = (5^{\ell_1 - 1} - 1)/(5 - 1). \tag{7.68}$$

Since $(\ell_1 - 1)/2$ is odd, it follows that $2 | S$, $2^2 \nmid S$. From (7.65)-(7.67), we therefore conclude that

$$2 | x_1, \quad 4 | x_{1i}, \quad 4 | x'_{1\nu}. \tag{7.69}$$

Referring to the relation (2.19), we have

$$1 = (\zeta_8^{x_i})^{\phi_1 - 1}(\zeta_8^{-x_{1i}})^{\rho-1}(\zeta_8^{-x_1})^{\phi_i - 1} = \zeta_8^{4x_i \pm 2x_{1i} - 4x_1}, \ (i \neq 1), \quad (7.70)$$

$$1 = (\zeta_8^{y_\nu})^{\phi_1 - 1}(\zeta_8^{-x'_{1\nu}})^{\rho-1}(\zeta_8^{-x_1})^{\xi_\nu - 1} = \zeta_8^{4y_\nu \pm 2x'_{1\nu}}, \quad (7.71)$$

whence by (7.69),

$$2|x_i \quad (1 \leqq i \leqq h), \qquad 2|y_\nu \quad (1 \leqq \nu \leqq t). \quad (7.72)$$

Referring to (2.19), we also have

$$1 = (\zeta_8^{x_{ij}})^{\rho-1}(\zeta_8^{-x_j})^{\phi_i - 1}(\zeta_8^{x_i})^{\phi_j - 1} = \zeta_8^{\pm 2x_{ij} - 4x_j + 4x_i},$$

$$1 = (\zeta_8^{x'_{i\nu}})^{\rho-1}(\zeta_8^{-y_\nu})^{\phi_i - 1}(\zeta_8^{x_i})^{\xi_\nu - 1} = \zeta_8^{\pm 2x'_{i\nu} - 4y_\nu},$$

$$1 = (\zeta_8^{y_{\nu\mu}})^{\rho-1}(\zeta_8^{-y_\mu})^{\xi_\nu - 1}(\zeta_8^{y_\nu})^{\xi_\mu - 1} = \zeta_8^{\pm 2y_{\nu\mu}},$$

and hence, by (7.72)

$$4|x_{ij}, \qquad 4|x'_{i\nu}, \qquad 4|y_{\nu\mu}, \quad (7.73)$$

where $1 \leqq i, j \leqq h$ $(i \neq j)$ and $1 \leqq \nu, \mu \leqq t$ $(\nu \neq \mu)$. Thus we conclude that

$$u_{\phi_i} u_{\phi_j} = \pm u_{\phi_j} u_{\phi_i}, \quad u_{\phi_i} u_{\xi_\nu} = \pm u_{\xi_\nu} u_{\phi_i}, \quad u_{\xi_\nu} u_{\xi_\mu} = \pm u_{\xi_\mu} u_{\xi_\nu}. \quad (7.74)$$

Referring to the relation (2.18), we have

$$(-1)^{z_j} = (\zeta_4^{z_j})^{\rho-1} = (\zeta_8^{x_j})^{1+\phi_j+\cdots+\phi_j^{\ell_j-2}} = \zeta_8^{x_j S_j}, \tag{7.75}$$

$$S_j = 1 + 5 + \cdots + 5^{\ell_j-2} = (5^{\ell_j-1} - 1)/(5 - 1), \tag{7.76}$$

$$\zeta_4^{\pm z'_\nu} = (\zeta_8^{z'_\nu})^{\rho-1} = \zeta_8^{y_\nu(1+\xi_\nu+\cdots+\xi_\nu^{p_\nu-2})} = \zeta_8^{y_\nu(p_\nu-1)}. \tag{7.77}$$

$(\zeta_8^2 = \zeta_4)$. For each $j = n + 1, \cdots, h$, $\ell_j \equiv 1 \pmod 4$, and so $4|S_j$. Since by (7.72), $2|x_j$, it follows from (7.75) that

$$2|z_j \quad \text{for} \quad n < j \leq h. \tag{7.78}$$

By (7.77) we have $z'_\nu \equiv \pm(p_\nu - 1)y'_\nu \pmod 4$, $y'_\nu = y_\nu/2$, where by (7.72), y'_ν is an integer. So

$$z'_\nu = \pm(p_\nu - 1)y'_\nu + 4z \tag{7.79}$$

for some integer z.

We will prove $\mathrm{inv}_{\ell_j}(B) = 0$ for $j = n + 1, \cdots, h$. The inertia group of ℓ_j in L/k is $\langle\phi_j^2\rangle$. Since $\ell_j \equiv 1$ or $5 \pmod 8$, and $\zeta_8^\rho = \zeta_8^{-1}$ or ζ_8^3, it follows that any Frobenius automorphism η of ℓ_j in L/k is of the form: $\eta = \phi_1^{a_1}\cdots\phi_h^{a_h}\xi_1^{a'_1}\cdots\xi_t^{a'_t}$. By (7.74), $u_{\phi_j}^2$ commutes with u_{ϕ_i} $(1 \leq i \leq h)$ and u_{ξ_ν} $(1 \leq \nu \leq t)$. Hence $\beta'(\phi_j^2, \eta)/\beta'(\eta, \phi_j^2) = 1$. Since by (7.78),

$$u_{\phi_j}^{\ell_j-1} = \zeta_4^{z_j} = \pm 1,$$

the same argument as in Case I yields that the ℓ_j-local index of B is equal to 1.

Next we will prove that if p_ν is inertial in k/Q, then $\mathrm{inv}_{p_\nu}(B) = 0$. The inertia group of p_ν in L/k is $\langle \xi_\nu \rangle$. Let a_i $(1 \leq i \leq h)$ and a'_μ $(1 \leq \mu \leq t;\ \mu \neq \nu)$ be the even numbers determined by (7.42), (7.43). Since $p_\nu^2 \equiv 1 \pmod 8$, we see easily that

$$\eta = \prod_i \phi_i^{a_i} \prod_{\mu \neq \nu} \xi_\mu^{a'_\mu}$$

is a Frobenius automorphism of p_ν in L/k. By (7.74), u_{ξ_ν} commutes with $u_{\phi_i}^2$, $u_{\xi_\mu}^2$ and hence also commutes with $u_{\phi_i}^{a_i}$, $u_{\xi_\mu}^{a'_\mu}$. It follows that $\beta'(\xi_\nu, \eta)/\beta'(\eta, \xi_\nu) = 1$. By (7.79), we have

$$u_{\xi_\nu}^{p_\nu - 1} = \zeta_8^{z'_\nu} = \zeta_8^{\pm(p_\nu-1)y'_\nu + 4z} = \zeta_4^{\pm y'_\nu (p_\nu-1)/2 + 2z} .$$

Thus, by the same argument as in Case I, we conclude that the p_ν-local index of B is equal to 1.

The proof of Theorem 7.14 is completed.

<u>Remark 7.15</u>. It is interesting that if there exists no prime $\ell \equiv 3 \pmod 4$ dividing m then $S(Q(\sqrt{m})) = M(Q(\sqrt{m}))$, and if there exists a prime $\ell \equiv 3 \pmod 4$ dividing m then $S(Q(\sqrt{m})) = S(Q) \otimes_Q Q(\sqrt{m})$.

Using the techniques of the proof of Theorem 7.14, we can also determine the Schur subgroup of the maximal real subfield $Q(\zeta_n + \zeta_n^{-1})$ of $Q(\zeta_n)$, where n is a natural number divisible by at least two distinct primes.

<u>Theorem 7.16</u> (Yamada [49]). Let n be a natural number such that $n = 2^a h$, $a \geq 2$, $h > 1$, $(2, h) = 1$. Let $k = Q(\zeta_n + \zeta_n^{-1})$. Then the Schur subgroup $S(k)$ consists of the classes of $Br(k)$ which have uniformly distributed invariants with values 0 or $\frac{1}{2}$ such that for a finite rational prime p, the p-local invariant is 0 whenever $p|n$, and whenever $p \nmid n$, f is even, $p^{f/2} \not\equiv -1 \pmod{n}$ and $p^{f/2} \equiv \pm 1 \pmod{2^a}$, where f is the smallest positive integer such that $p^f \equiv 1 \pmod{n}$.

<u>Theorem 7.17</u> (Yamada [49]). Let n be an odd integer divisible by at least two distinct primes. Let $k = Q(\zeta_n + \zeta_n^{-1})$. Then the elements of $S(k)$ are those classes of $Br(k)$ which have uniformly distributed invariants with values 0 or $\frac{1}{2}$ such that the p-local invariant is 0 whenever $p|n$, and whenever $p \nmid n$, f is even and $p^{f/2} \not\equiv -1 \pmod{n}$, where f is the smallest positive integer such that $p^f \equiv 1 \pmod{n}$.

For the proofs of Theorems 7.16, 7.17, see Theorems 1 and 1' of [49].

Chapter 8. THE SCHUR SUBGROUP OF AN IMAGINARY FIELD

Let k be a cyclotomic extension of Q. For a prime p, we write $S(k)_p$ for the p-primary part of $S(k)$, i.e., the subgroup of $S(k)$ consisting of all classes of p-power order. Throughout this chapter, k is assumed to be imaginary. Let k_0 denote the maximal real subfield of k ($[k : k_0] = 2$). Let A be a Schur algebra over k. We have proved in Proposition 1.10 that A has an involution of the second kind (over k_0). We will quote a theorem on involutorial algebras:

Theorem 8.1 (Theorem 21, p.161 of [1]). Let $K \supset E$ be fields with $[K : E] = 2$. A quaternion division algebra A central over K has an involution over E if and only if $A = B \otimes_E K$, where B is a quaternion division algebra central over E.

Let $K \supset E$ be as in Theorem 8.1. Let D be a division algebra central over K and n a positive integer. Then the central simple algebra $M_n(D)$ over K is involutorial over E if and only if D is involutorial over E (Theorem 13, p.156 of [1].)

Now we proceed to determine the Schur subgroup of an imaginary field.

Proposition 8.2 (Benard-Schacher [5]). Let k be a

cyclotomic extension of Q. Suppose k is not real and $G(k/Q)$ has a cyclic Sylow 2-subgroup. Then the elements of $S(k)$ of order 2 are those classes induced from $S(Q)$, i.e., $S(Q) \otimes_Q k$.

Proof. Suppose $[A] \in S(k)$ has order 2. By Theorem 7.2, $S(Q)$ consists of all classes of quaternion algebras central over Q. We see easily that in order to prove $[A] = [B \otimes_Q k]$ for some $[B] \in S(Q)$, it suffices to show that A has zero p-local invariant whenever $[Q_p k : Q_p]$ is divisible by 2.

Suppose p is such that A has non-zero p-local invariant. Let k_0 be the maximal real subfield of k. By Theorem 8.1, $A \sim B \otimes_{k_0} k$ where B is a quaternion division algebra central over k_0. Let $\mathfrak{p}$ be a prime of k dividing p and put $\mathfrak{p}' = \mathfrak{p} \cap k_0$. Then

$$d \cdot \mathrm{inv}_{\mathfrak{p}'}(B) \equiv \mathrm{inv}_{\mathfrak{p}}(A) = \frac{1}{2} \pmod{Z}$$

where $d = [Q_p k : Q_p k_0]$. Thus $d = 1$. Since $k \neq k_0$, it follows that the unique involution ι of $G(k/Q)$ does not fix the prime $\mathfrak{p}$. Hence the decomposition group $H = \{\sigma \in G(k/Q);\ \mathfrak{p}^\sigma = \mathfrak{p}\}$ has odd order. Since $[Q_p k : Q_p] = |H|$, $[Q_p k : Q_p]$ is not divisible by 2. #

Theorem 8.3 (Benard-Schacher [5]). Let k be an imaginary quadratic field with $\sqrt{-1} \notin k$ and $\sqrt{-3} \notin k$. Then $S(k) = S(Q) \otimes_Q k$. That is, $S(k)$ consists of the classes of

Br(k) which have uniformly distributed invariants with values 0 or $\frac{1}{2}$ such that the p-local invariant is 0 whenever p does not split in k.

Proof. By Theorem 6.1, S(k) consists only of elements of order 2. Then Proposition 8.2 yields that $S(k) = S(Q) \otimes_Q k$. The last statement follows from this, because S(Q) consists of all classes of quaternion algebras central over Q. #

Remark 8.4. Fields-Herstein [24] hit on the notion of using Theorem 8.1 for the theory of Schur subgroup.

Next we consider the case that $k = Q(\zeta_{\ell^n})$, where ℓ is a prime. This includes the cases of the imaginary quadratic fields $Q(\sqrt{-1})$ and $Q(\sqrt{-3})$ which were excluded in Theorem 8.3. Since S(Q) has already been determined, we assume $\ell^n > 2$.

Let p be an odd prime with $p \equiv 1 \pmod{\ell}$. Suppose

$$p = 1 + \ell^c d \quad (c \geqq 1) \quad \text{with} \quad (\ell, d) = 1. \qquad (8.1)$$

Let t be an integer which is a primitive root modulo p. Let ζ be a primitive $p\ell^n$-th root of unity. Then $\zeta^p = \zeta_{\ell^n}$ is a primitive ℓ^n-th root of unity. The Galois group $G = G(Q(\zeta)/k)$, $k = Q(\zeta_{\ell^n})$, is cyclic of order $p - 1$: $G = \langle\sigma\rangle$, $\zeta^\sigma = \zeta^v$, where $v \equiv 1 \pmod{\ell^n}$, $v \equiv t \pmod{p}$. Consider the following cyclic algebra A_p over k, which is also a cyclotomic algebra:

$$A_p = (\zeta_{\ell^n}, Q(\zeta)/k, \sigma) = \sum_{i=0}^{p-2} Q(\zeta)u^i, \tag{8.2}$$

$$u^{p-1} = \zeta^p = \zeta_{\ell^n}, \qquad u\zeta u^{-1} = \zeta^\sigma. \tag{8.3}$$

For a rational prime y, $y\text{-ind}(A_p)$ denotes the y-local index of A_p.

<u>Lemma 8.5</u>. If y is a rational prime $\neq p$, then $y\text{-ind}(A_p) = 1$. If either $\ell \neq 2$ or $p \not\equiv -1 \pmod{\ell^n}$, then $p\text{-ind}(A_p) = \ell^s$, where $s = \text{Min}\{n, c\}$. If $\ell = 2$ and $p \equiv -1 \pmod{2^n}$, then $p\text{-ind}(A_p) = 1$.

<u>Proof</u>. If y is a prime number $\neq p, \ell$, then $y\text{-ind}(A_p) = 1$, because y is unramified in $Q(\zeta)/k$. If $y = \infty$, the rational infinite prime, then $\infty\text{-ind}(A_p) = 1$, because k is imaginary. If $y = \ell$, then Proposition 4.8 and Corollary 5.4 yield that $\ell\text{-ind}(A_p) = 1$, because if $\ell \neq 2$ then $\zeta_\ell \in k$, and if $\ell = 2$ then $\zeta_4 \in k$. This proves the first assertion.

The p-local index of A_p can be determined by the results of Appendix to Chapter 4. Let f be the smallest positive integer such that $p^f \equiv 1 \pmod{\ell^n}$. Write

$$\zeta_{\ell^n} = \zeta^v_{p^f-1}, \qquad v = (p^f - 1)/\ell^n. \tag{8.4}$$

Then the formulas (4.16), (4.17) imply that

$$p\text{-ind}(A_p) = (p - 1)/(v, p - 1). \tag{8.5}$$

It is clear that if $s = n$ then $f = 1$ and $(p - 1)/(v, p - 1) = \ell^n = \ell^s$.

Suppose that either $\ell \neq 2$ or $c > 1$, and that $s = c$. Then it is easy to see that $f = \ell^{n-c}$, $p^f - 1 \not\equiv 0 \pmod{\ell^{n+1}}$, and $(p - 1)/(v, p - 1) = \ell^c = \ell^s$.

Suppose that $\ell = 2$ and $c = 1$, i.e., $p = 1 + 2d$, $(2, d) = 1$. In this case, $s = c$. We furthermore assume $p \not\equiv -1 \pmod{2^n}$. Then $2^n \nmid p + 1$ and hence the following two cases only occur: (i) $2^n \| p^2 - 1$, (ii) $2^n \nmid p^2 - 1$. For the case (i), we have $f = 2$, $(p - 1)/(v, p - 1) = (p - 1)/((p^2 - 1)/2^n, p - 1) = 2 = 2^c = 2^s$. For the case (ii), write $p^2 - 1 = 2^t d'$. Then $3 \leqq t < n$, and hence $f = 2^{n-t+1}$, $2^n \| p^f - 1$. Consequently, $(p - 1)/(v, p - 1) = 2 = 2^s$.

Thus we have proved that $p\text{-ind}(A_p) = \ell^s$ except $\ell = 2$ and $p \equiv -1 \pmod{2^n}$. On the other hand, if $\ell = 2$ and $p \equiv -1 \pmod{2^n}$, then $f = 2$ and $p^2 - 1 \equiv 0 \pmod{2^{n+1}}$. Consequently, $p\text{-ind}(A_p) = (p - 1)/(v, p - 1) = (p - 1)/((p^2 - 1)/2^n, p - 1) = 1$. This completes the proof of Lemma 8.5.

<u>Theorem 8.6</u> (Benard-Schacher [5]). Let $k = Q(\zeta_{\ell^n})$ with $\ell^n > 2$. Then the ℓ-primary part $S(k)_\ell$ of $S(k)$ is generated by the classes $[A_p]$ given by (8.2), (8.3), where if $\ell \neq 2$ then p ranges over all primes such that $p \equiv 1 \pmod{\ell}$, and if $\ell = 2$ then p ranges over all odd primes such that $p \not\equiv -1 \pmod{2^n}$.

Proof. Let $[A] \in S(k)_\ell$. Let $\{p_1, p_2, \cdots, p_t\}$ be the distinct rational primes such that p_i-ind$(A) > 1$, and that $p_i \not\equiv -1 \pmod{2^n}$ if $\ell = 2$. P is a finite set and it contains no infinite primes since k is imaginary. By Proposition 4.8 and Corollary 5.4, $\ell \notin P$ since $\zeta_\ell \in k$ for $\ell \neq 2$ and $\zeta_4 \in k$ for $\ell = 2$.

Let $p \in P$ and suppose A has p-local index $\ell^a > 1$. By Theorem 4.3, $p = 1 + \ell^c d$ with $c \geqq a$. By Theorem 6.1, $\zeta_{\ell^a} \in k$ so $n \geqq a$. Set $s = \min\{c, n\}$. Let A_p be the cyclotomic algebra over k given by (8.2), (8.3) with p-local index ℓ^s and all other local indices equal to 1. Let $\mathfrak{p}$ be a prime of k dividing p such that $\mathrm{inv}_{\mathfrak{p}}(A_p) = 1/\ell^s$. Suppose $\mathrm{inv}_{\mathfrak{p}}(A) = \mu/\ell^a$. Set $h = \mu\ell^s/\ell^a$. Then by Theorem 6.1, $[A_p]^h$ and $[A]$ have equal p-local invariants, i.e., $\mathrm{inv}_{\mathfrak{p}'}([A_p]^h) = \mathrm{inv}_{\mathfrak{p}'}([A])$ for any $\mathfrak{p}'|p$.

For each $p_i \in P$, let $[A_{p_i}]^{h_i}$ be selected as in the preceding paragraph. If $\ell \neq 2$, then $[A]$ and $[A_{p_1}]^{h_1}\cdots[A_{p_t}]^{h_t}$ have equal invariants, so $[A] = [A_{p_1}]^{h_1}\cdots[A_{p_t}]^{h_t}$.

Suppose next that $\ell = 2$. Let A' be a Schur algebra over k such that $[A'] = [A][A_{p_1}]^{-h_1}\cdots[A_{p_t}]^{-h_t}$. For any rational prime $p' \not\equiv -1 \pmod{2^n}$, the p'-local index of A' equals 1. The index of A' is a power of 2. On the other hand, it is equal to the least common multiple of the p-local indices of A', where p ranges over all primes $p \equiv -1$

(mod 2^n). Since $2^2 \nmid p-1$, it follows from Theorem 4.4 that the index of A' is at most 2. Let k_0 denote the maximal real subfield of k, i.e., $k_0 = Q(\zeta_{2^n} + \zeta_{2^n}^{-1})$. If the index of A' would be equal to 2, then Theorem 8.1 yields that there exists a quaternion division algebra B central over k_0 such that $A' \sim B \otimes_{k_0} k$. There exists a prime $p \equiv -1 \pmod{2^n}$ with $p\text{-ind}(A') = 2$. But p is inertial in k/k_0, so $[kQ_p : k_0Q_p] = 2$ and $\mathrm{inv}_p(A') = [kQ_p : k_0Q_p]\cdot\mathrm{inv}_p(B) = 0$, contradiction. Thus the index of A' is equal to 1 and $[A] = [A_{p_1}]^{h_1}\cdots[A_{p_t}]^{h_t}$, proving the theorem.

<u>Corollary 8.7</u>. Let $k = Q(\zeta_{2^n})$, $n \geq 2$. Let $[A] \in S(k)$. Then for any prime $p \equiv -1 \pmod{2^n}$, the p-local index of A is equal to 1.

<u>Proof</u>. This is clear from Lemma 8.5 and Theorem 8.6.

<u>Remark 8.8</u>. The case $\ell = 2$, $p \equiv -1 \pmod{2^n}$ is exceptional, and is overlooked in Benard-Schacher [5].

<u>Theorem 8.9</u> (Benard-Schacher [5]). Let $k = Q(\zeta_{\ell^n})$ with $\ell^n > 2$. Then $S(k) = S(k)_\ell \times S(k)_2$ unless $\ell = 2$, in which case $S(k) = S(k)_\ell$. $S(k)_\ell$ is the ℓ-primary part of $S(k)$ and is determined by Theorem 8.6. If $\ell \neq 2$, then $S(k)_2 = S(Q) \otimes_Q k$, i.e., $S(k)_2$ consists of those classes of $Br(k)$ which have uniformly distributed invariants with values

0 or $\frac{1}{2}$ such that the p-local invariants are 0 whenever $p = \ell, \infty$, and whenever $p \neq \ell, \infty$ and f is even, f being the least positive integer such that $p^f \equiv 1 \pmod{\ell^n}$.

Proof. When $\ell = 2$, $S(k)$ contains no elements of odd order so it is characterized by Theorem 8.6. Suppose ℓ is odd. By Theorem 6.1, $S(k) = S(k)_\ell \times S(k)_2$ where $S(k)_2$ consists of the elements of order 2. By Proposition 8.2, elements of order 2 are induced from $S(Q)$. They have zero p-local invariants whenever $f = [Q_p k : Q_p]$ is divisible by 2. Note that $[Q_\ell k : Q_\ell] = [k : Q]$ is divisible by 2. For a prime $p \neq \ell$, we know that f is the smallest integer such that $p^f \equiv 1 \pmod{\ell^n}$. The theorem now follows.

Theorem 8.10 ([48]). Let ℓ be a prime. Let k be an imaginary subfield of $Q(\zeta_{\ell^n})$ other than $Q(\zeta_{\ell^c})$ $(c = 1, 2, \cdots, n)$. Then $S(k) = S(Q) \otimes_Q k$.

Proof. If $\ell = 2$, then by (7.12)-(7.14) and Proposition 7.5, we see that k/Q is cyclic and the roots of unity in k are ± 1. If $\ell \neq 2$, these statements clearly hold for k. The theorem now follows from Theorem 6.1 and Proposition 8.2.

Remark 8.11. In view of Theorem 7.4, Theorem 8.9 and Theorem 8.10, the Schur subgroup $S(k)$ of any arbitrary subfield k of $Q(\zeta_{\ell^n})$ is completely determined.

Let m be an arbitrary positive integer > 2, and ζ_m a primitive m-th root of unity. If m is odd then $Q(\zeta_m) = Q(\zeta_{2m})$. Hence we may assume that m is either odd or divisible by 4.

Theorem 8.12 (Janusz [32]). Let m be a positive integer which is either odd or divisible by 4. Let $m = \ell^a m'$ with ℓ a prime not dividing m'. Then

$$S(Q(\zeta_m))_\ell = S(Q(\zeta_{\ell^a}))_\ell \otimes Q(\zeta_m). \tag{8.6}$$

Remark 8.13. If m is odd, then (8.6) implies, as a special case,

$$S(Q(\zeta_m))_2 = S(Q) \otimes_Q Q(\zeta_m). \tag{8.7}$$

Proof of Theorem 8.12. By using Theorem 7.9, we can give a simple proof of the theorem. Put $k = Q(\zeta_m)$. Suppose first that m is odd. Let B be a cyclotomic algebra over k. By Theorem 7.9, B may be assumed to be of the form:

$$B = (\beta, k(\zeta_b)), \quad b = 4p_1p_2\cdots p_s, \tag{8.8}$$

where $p_1, p_2, \cdots, p_s$ are distinct odd primes not dividing m. If ℓ is odd and $[B] \in S(k)_\ell$, then we may assume $\beta(\sigma, \tau) \in \langle\zeta_{\ell^a}\rangle$ for all $\sigma, \tau \in G(k(\zeta_b)/k)$. Put $k' = Q(\zeta_{\ell^a})$. It is clear that $k'(\zeta_b) \cap k = k'$ and $k'(\zeta_b)\cdot k = k(\zeta_b)$.

Consequently, $G(k(\zeta_b)/k) \simeq G(k'(\zeta_b)/k')$ and $B \simeq (\beta, k'(\zeta_b)/k') \otimes_{k'} k$. This implies $[B] \in S(k')_\ell \otimes_{k'} k$. If $\ell = 2$ and $[B] \in S(k)_2$, then we may assume $\beta(\sigma, \tau) \in \langle\zeta_4\rangle$ for all $\sigma, \tau \in G(k(\zeta_b)/k)$. Since $Q(\zeta_b) \cap k = Q$ and $Q(\zeta_b)\cdot k = k(\zeta_b)$, it follows that $B \simeq (\beta, Q(\zeta_b)/Q) \otimes_Q k$, and $[B] \in S(Q) \otimes_Q k$.

Suppose next that m is divisible by 4. Then by Theorem 7.9, a cyclotomic algebra B over $k = Q(\zeta_m)$ may be assumed to be of the form:

$$B = (\beta, k(\zeta_b)/k), \qquad b = p_1 p_2 \cdots p_s, \tag{8.9}$$

where $p_1, p_2, \cdots, p_s$ are distinct odd primes not dividing m. If $[B] \in S(k)_\ell$, we may assume $\beta(\sigma, \tau) \in \langle\zeta_{\ell^a}\rangle$ for $\sigma, \tau \in G(k(\zeta_b)/k)$. By the same argument as before, we conclude $[B] \in S(Q(\zeta_{\ell^a}))_\ell \otimes k$. #

Now we consider a field k contained in some cyclotomic field $Q(\zeta_n)$, where n is either odd or divisible by 4. Also let m be the order of the group of roots of unity in k. For a field L containing k we write $S(L/k)_\ell$ for the subgroup of $S(k)_\ell$ consisting of algebra classes split by L.

Proposition 8.14 (Janusz [32]). Let $m = \ell^a m'$ and $n = \ell^b n'$ with ℓ a prime not dividing $m'n'$. Suppose $a \geq b$. Then $S(k)_\ell = \{S(Q(\zeta_{\ell^b}))_\ell \otimes k\}\cdot S(Q(\zeta_n)/k)_\ell$.

Proof. For a class $[\Delta]$ in $S(k)_\ell$ we have a corresponding class $[\Delta \otimes Q(\zeta_n)]$ in $S(Q(\zeta_n))_\ell$. By Theorem 8.12 there is a class $[\Gamma]$ in $S(Q(\zeta_{\ell^b}))_\ell$ such that

$$[\Delta \otimes Q(\zeta_n)] = [\Gamma \otimes Q(\zeta_n)]. \tag{8.10}$$

(The tensor products are taken over the center of the algebra in the left factor of each product.) Our assumptions insure $Q(\zeta_{\ell^b}) \subset k$ so $\Gamma \otimes k$ is central simple over k. It now follows from (8.10) that $[\Delta]\cdot[\Gamma \otimes k]^{-1} \in S(Q(\zeta_n)/k)_\ell$ and so $[\Delta]$ is in the subgroup mentioned in the statement of the proposition and the proof is complete.

Corollary 8.15. With the same notation and assumption as in the proposition, suppose ℓ does not divide $[Q(\zeta_n) : k]$. Then $S(k)_\ell = S(Q(\zeta_{\ell^b}))_\ell \otimes k$.

Proof. The exponent of an algebra split by $Q(\zeta_n)$ divides $[Q(\zeta_n) : k]$. When this dimension is not divisible by ℓ we have $S(Q(\zeta_n)/k)_\ell = 1$ and the result follows.

Corollary 8.16. Let k be an abelian extension of Q which contains no roots of unity except ± 1. Suppose $k \subset Q(\zeta_n)$ with n odd. Then $S(k) = \{S(Q) \otimes k\}\cdot S(Q(\zeta_n)/k)$.

Proof. By the root of unity theorem (Theorem 6.1) we have $S(k) = S(k)_2$ so the proposition gives the result.

Chapter 9. SOME THEOREMS FOR A SCHUR ALGEBRA

Theorem 9.1 (Fein-Yamada [20]). Let χ be a complex irreducible character of a finite group G, and let $m = m_Q(\chi)$ denote the Schur index of χ over the rationals Q. Then m divides the exponent of G and m^2 divides the order of G. Let p be a prime. If p^r divides m and either p equals 2 or p is an odd prime such that the Sylow p-subgroups of G are abelian, then p^{r+1} divides the exponent of G.

Proof. Let p be a prime, $p^r | m$, $p^{r+1} \nmid m$, $r \geq 1$. Let n be the exponent of G and let L be the subfield of $Q(\zeta_n)$, ζ_n a primitive n-th root of unity, such that $L \supset Q(\chi)$, $p \nmid [L : Q(\chi)]$, and $[Q(\zeta_n) : L]$ is a power of p. By the Brauer-Witt theorem there is an L-elementary subgroup H of G with respect to p and an irreducible complex character θ of H with the following properties:

(1) there is a normal subgroup N of H and a linear character ψ of N such that $\theta = \psi^H$;

(2) $H/N \simeq \mathcal{G}(L(\psi)/L)$;

(3) $L(\theta) = L$;

(4) $m_L(\theta) = p^r$;

(5) for every $h \in H$ there is a $\tau(h) \in \mathcal{G}(L(\psi)/L)$ such that $\psi(hnh^{-1}) = \tau(h)(\psi(n))$ for all $n \in N$;

(6) $A(\theta, L) \underset{L}{\simeq} (\beta, L(\psi)/L)$ where, if D is a complete set of coset representatives of N in H $(1 \in D)$ with $hh' = n(h, h')h''$ for $h, h', h'' \in D$, $n(h, h') \in N$, then

$\beta(\tau(h), \tau(h')) = \psi(n(h, h'))$.

Since the index of $A(\theta, L)$ is p^r, it follows from (2) and (6) that p^r divides $[L(\psi) : L] = [H : N]$. Thus we need only show that N has exponent divisible by p^r and we will have proved the first two assertions of the theorem.

The exponent of the cyclotomic algebra $(\beta, L(\psi)/L)$ is equal to p^r. Let $\langle\beta\rangle$ denote the subgroup of $L(\psi)^*$ generated by the values of β and suppose $\langle\beta\rangle$ has order $c = p^d t$, $(p, t) = 1$. We have $[(\beta, L(\psi)/L)]^c = [(\beta^c, L(\psi)/L)] = [(1, L(\psi)/L)] = 1$ in $Br(L)$ so $d \geq r$. There are $h, h' \in D$ with the p-part of the order of $\beta(\tau(h), \tau(h'))$, $|\beta(\tau(h), \tau(h'))|_p$, equal to p^d. Since $\beta(\tau(h), \tau(h')) = \psi(n(h, h'))$, $n(h, h')$ has order divisible by p^d. In particular, p^r divides the exponent of N.

Let $H/N = \langle Nh_1\rangle \times \cdots \times \langle Nh_s\rangle$ where Nh_i has order s_i in H/N (s_i a power of p for all i). Set

$$D = \{h_1^{a_1} \cdots h_s^{a_s};\quad 0 \leq a_i < s_i\}.$$

Since H is L-elementary with respect to p, H is the semidirect product of a Sylow p-subgroup P of H and a cyclic normal p'-group. We may clearly assume that $D \subset P$ and $\{n(h, h');\ h, h' \in D\} \subset P$. Let $h, h' \in D$ with $|\psi(n(h, h'))|_p = p^d$. It is an easy verification that $n(h, h')$ is a product of conjugates (under elements of H) of elements of the form

$n(h_i, h_j)$ with $i > j$ and $n(h_i^{s_i - 1}, h_i)$. Thus, we see from (5) that we either have

$$|\psi(n(h_i^{s_i - 1}, h_i))|_p = p^d \tag{9.1}$$

for some i or

$$|\psi(n(h_i, h_j))|_p = p^d \tag{9.2}$$

for some i, j with $i > j$. If (9.1) holds, then $h_i^{s_i} = n(h_i^{s_i - 1}, h_i)$ has order divisible by p^d and so p^{r+1} divides the exponent of H in this case. If p is odd, then P is abelian. Since $h_i h_j = n(h_i, h_j) h_j h_i$ we conclude, for p odd, that $n(h_i, h_j) = 1$ for $i > j$. Thus, if p is odd we must have $|\psi(n(h_i^{s_i - 1}, h_i))|_p = p^d$ for some i, proving the theorem in that case.

We may, therefore, assume that $p = 2$, $d = r$, $|\psi(n(h_i, h_j))|_2 = 2^r$, and the exponent of H/N divides 2^r. Since $n(h_i, h_j) \in P$, $\psi(n(h_i, h_j)) = \zeta$, a primitive 2^r-th root of unity. By Theorem 6.1, $\zeta \in L$ and so, if $\tau \in G(L(\psi)/L)$, $\tau(\zeta) = \zeta$. We have

$$(h_j h_i)^{2^r} = h_j h_i h_j h_i \cdots h_j h_i = n_1 n_2 \cdots n_{2^r - 1} h_j^{2^r} h_i^{2^r},$$

where n_t is a product of t conjugates of $n(h_i, h_j)$. Using (5) we see that

$$\psi((h_j h_i)^{2^r}) = \zeta^b \psi(h_j^{2^r})\psi(h_i^{2^r})$$

where $b = 2^{r-1}(2^r - 1)$. Thus, $\zeta^b = -1$. In particular, one of $(h_j h_i)^{2^r}$, $h_j^{2^r}$, $h_i^{2^r}$, has even order, proving the theorem.

We have the following immediate consequence.

Corollary 9.2. Let χ be a complex irreducible character of a finite group G and suppose every Sylow subgroup of G is elementary abelian. Then $m_Q(\chi) = 1$.

Remark 9.3. If p is an odd prime and the Sylow p-subgroups of G are non-abelian, then it is not necessary that p^{r+1} divides the exponent of G. Indeed, we can give an example (actually, one example for each odd prime p) of a group G and an irreducible character χ of G such that $m_Q(\chi) = p$ = p-part of the exponent of G. (See [20].) For every prime p, examples can also be given where $m_Q(\chi) = p$ but p^3 does not divide the order of G.

Theorem 9.4. (Goldschmidt and Isaacs [28]). Let G be a finite group with exponent n. Let k be a field of characteristic 0 and ζ_n a primitive n-th root of unity. Suppose for some prime p, that a Sylow p-subgroup P of the Galois group $G(k(\zeta_n)/k)$ is cyclic. If 2 divides the order of P, assume also that $\sqrt{-1} \in k$. Then the Schur index over k of every irreducible character of G is relatively prime to p.

Proof. Assume, by way of contradiction, that some irreducible character χ of G has Schur index divisible by p. Let L be the subfield of $k(\zeta_n)$ such that $L \supset k(\chi)$, $[k(\zeta_n) : L]$ is a power of p, and $[L : k(\chi)] \not\equiv 0 \pmod p$. Our assumption implies that $G(k(\zeta_n)/L)$ is cyclic. By the Brauer-Witt Theorem, there is a group H which is a section of G, and an irreducible character θ of H such that there exists a cyclic, normal subgroup N of H and a faithful linear character ψ of N with the following properties: (i) $\theta = \psi^H$, (ii) H/N is an abelian p-group and $H/N \simeq G(L(\psi)/L)$, (iii) $L(\theta) = L$, (iv) $A(\theta, L) \simeq (\beta, L(\psi)/L)$, (the factor set β depends on a complete set of representatives of N in H), and (v) p divides the index of the cyclotomic algebra in (iv). (See also Remark 3.12.)

Set $r = |N|$ and $E = L(\zeta_r)$. Then $E = L(\psi)$. Since N is a cyclic section of G, we have $r|n$, and $G(E/L)$ is a section of $G(k(\zeta_n)/L)$. By (ii), the p-group H/N is isomorphic to $G(E/L)$ and thus is cyclic. It is easy to see that there is an element y of H such that $H/N = \langle Ny \rangle$ and the order of y is a power of p. Put $p^b = |\langle y \rangle|$ and $p^a = |\langle y \rangle \cap N|$. Then $\{1, y, y^2, \cdots, y^{p^{b-a}-1}\}$ is a complete set of coset representatives of N in H, and the cyclotomic algebra in (iv) is the cyclic algebra

$$(\delta,\ E/L,\ \sigma), \quad \delta = \psi(y^{p^{b-a}}), \quad \langle\sigma\rangle = G(E/L). \qquad (9.3)$$

Since ψ is a faithful linear character of N and $y^{p^{b-a}}$ (ε N) has order p^a, it follows that $\delta = \psi(y^{p^{b-a}})$ is a primitive p^a-th root of unity. In particular, L contains a primitive p^a-th root of unity ζ_{p^a}. If $p^a = 1$, then $\delta = 1$ and $(\delta, E/L, \sigma) \sim L$, contradiction. Hence we have $a > 0$. We argue that E contains a primitive p^b-th root of unity ζ_{p^b}.

Since $p^b | n$, we see that E and $L(\zeta_{p^b})$ are intermediate fields lying between L and $k(\zeta_n)$. Moreover, $[E : L] = |H/N| = p^{b-a}$, and since $\zeta_{p^a} \in L$ for $a > 0$, we have

$$[L(\zeta_{p^b}) : L] = p^e \leqq p^{b-a}.$$

Now $G(k(\zeta_n)/L)$ is a cyclic p-group. It follows that the fields between L and $k(\zeta_n)$ are linearly ordered, and we conclude that $L(\zeta_{p^b}) \subset E$ as asserted.

Next we claim that $N_{E/L}(\zeta_{p^b})$ is a primitive p^a-th root of unity. Let $\zeta_{p^b}^\sigma = \zeta_{p^b}^z$ for some integer z. Then

$$N_{E/L}(\zeta_{p^b}) = \prod_{i=0}^{p^{b-a}-1} \zeta_{p^b}^{z^i} = \zeta_{p^b}^h$$

where $h = (z^{p^{b-a}} - 1)/(z - 1)$. Thus it suffices to show that p^{b-a} is the exact power of p dividing h. For $p \neq 2$,

$\zeta_p \in L$ (indeed, $\zeta_{p^a} \in L$ with $a > 0$) and $\zeta_p = \zeta_p^\sigma = \zeta_p^z$, so $z \equiv 1 \pmod p$. For $p = 2$, $\zeta_4 \in L$ and $\zeta_4 = \zeta_4^\sigma = \zeta_4^z$, so $z \equiv 1 \pmod 4$. It follows easily that for any prime p, p^{b-a} is the exact power of p dividing h.

We have now proved that the image of $N_{E/L}$ contains a primitive p^a-th root of unity. On the other hand, we have seen that the number δ in (9.3) is a primitive p^a-th root of unity. Thus, $(\delta, E/L, \sigma) \sim L$. This is a contradiction, because by (v), the index of the above cyclic algebra is divisible by p. The proof of the theorem is completed.

Remark 9.5. For $p = 2$, Fein [19] strengthened Theorem 9.4 by replacing the assumption $\zeta_4 \in k$ with the assumption $\sqrt{-1} = a^2 + b^2$ for $a, b \in k$.

Corollary 9.6 (Fong [25], Yamada [52]). Let G be a finite group of exponent n. Let p be a prime and let $n = p^c b$, where $(p, b) = 1$. Let k be a field of characteristic 0 which contains a primitive b-th root of unity. If $p = 2$ assume also that $\sqrt{-1} \in k$. Then $m_k(\chi) = 1$ for every irreducible character χ of G.

Proof. It is clear that $k(\zeta_n)$ is a cyclic extension of k. Since $k(\zeta_n)$ is a splitting field for every irreducible character χ of G, it follows that $m_k(\chi)$ divides

$[k(\zeta_n) : k]$. So the result follows from Theorem 9.4 immediately.

Remark 9.7. In Fong [25], the above n denoted the order of G.

Corollary 9.8 (Brauer [12], Roquette [33], Solomon [36]). Let G be a nilpotent group and χ an irreducible character of G. If $|G|$ is odd then $m_Q(\chi) = 1$, and if $|G|$ is even then $m_{Q(\sqrt{-1})}(\chi) = 1$.

Proof. This is clear from Corollary 9.6 and Proposition 1.9.

Corollary 9.9 (Solomon [36]). Let G be a finite group of exponent n. Let $p_1, p_2, \cdots, p_s$ be the primes dividing n. Let k be a field of characteristic 0 which contains a primitive $p_1 p_2 \cdots p_s$-th root of unity. If n is even, assume also that $\sqrt{-1} \in k$. Then $m_k(\chi) = 1$ for any irreducible character χ of G. Thus in particular $m_Q(\chi) \mid 2(p_1 - 1)\cdots(p_s - 1)$ and if $|G|$ is odd then $m_Q(\chi) \mid (p_1 - 1)\cdots(p_s - 1)$.

Proof. It is easy to see that if a prime p divides $|G(k(\zeta_n)/k)|$, then $p \in \{p_1, p_2, \cdots, p_s\}$ and the Sylow p-subgroup of $G(k(\zeta_n)/k)$ is cyclic. So the result follows from Theorem 9.4.

Theorem 9.10 ([52]). Let p be a prime number and Q_p the rational p-adic field. Let G be a finite group of exponent n. Suppose that $p \nmid n$ for $p \neq 2$, and that $4 \nmid n$ for $p = 2$.

Then $m_{Q_p}(\chi) = 1$ for every irreducible character χ of G.

Proof. Set $k = Q_p(\chi)$. Then $m_k(\chi) = m_{Q_p}(\chi)$. Let ℓ be a prime. By the Brauer-Witt theorem, the ℓ-part of $m_k(\chi)$ is equal to the index of some cyclotomic algebra of the form $(\beta, L(\psi)/L)$, where $Q_p \subset k \subset L \subset L(\psi) \subset Q_p(\zeta_n)$. It follows from the assumption that the extension $Q_p(\zeta_n)/Q_p$ is unramified. Hence $(\beta, L(\psi)/L) \sim L$. As ℓ is an arbitrary prime, we conclude that $m_k(\chi) = 1$.

NOTATION AND TERMINOLOGY

Z and Q are, respectively, the ring of integers and the field of rational numbers.

Let n be a positive integer. ζ_n denotes a primitive n-th root of unity. $Z \bmod^{\times} n$ is the multiplicative group of integers modulo n.

Let $z \in Z$ and let p be a prime number. If p^m is the exact power of p dividing z, then we call p^m the p-part of z and write $p^m \| z$.

If A is a ring with 1, then $M_n(A)$ is the ring of $n \times n$ matrices with coefficients in A. A^* is the group of invertible elements of A.

All fields are assumed to be of characteristic 0. Let k be a field. We say that a field K is a cyclotomic extension of k, only if there is a root of unity ζ and an element α of the cyclotomic field $Q(\zeta)$ such that $K = k(\alpha)$.

$G(K/k)$ is the Galois group of K over k.

For $\sigma \in G(K/k)$ and $x \in K$, both $\sigma(x)$ and x^σ denote the image of x by σ.

$N_{K/k}$ is the norm of K over k.

The 2-cohomology group $H^2(G(K/k), K^*)$ is, as usual, denoted by $H^2(K/k)$.

Let K and k be cyclotomic extensions of Q such that $K \supset k$. Let p be a rational prime and $\mathcal{P}$ (resp. $\mathfrak{p}$) a prime of K

(resp. k) lying above p. Then K^{P}/k_{p} represents the isomorphy type of the completion of K/k for $P|p$. We refer the ramification index (resp. the residue class degree) of P from k to K as the ramification index (resp. the resicue class degree) of p in K/k. If T (resp. ϕ) is the inertia group (resp. a Frobenius automorphism) of P with respect to the extension K/k, then we say that T (resp. ϕ) is the inertia group (resp. a Frobenius automorphism) of p in K/k, etc.

Let A and B be central simple algebras. If A is similar to B, we write $A \sim B$.

If k is a finite extension of Q_p, the rational p-adic numbers, then $inv_k(A)$ is the (Hasse) invariant of A.

If k is a finite extension of Q and p a prime of k, then $inv_p(A)$ is the invariant of A at p.

All groups are assumed to be finite. Let G be a group.

$|G|$ is the cardinality of G.

By an irreducible character χ of G, we mean an absolutely irreducible one.

$m_k(\chi)$ is the Schur index of χ over k.

$k(\chi)$ is a field obtained from k by adjunction of all values $\chi(g)$, $g \in G$.

For $\sigma \in G(k(\chi)/k)$, χ^{σ} is the character of G defined by $\chi^{\sigma}(g) = \sigma(\chi(g))$ for all $g \in G$.

If χ and ψ are class functions on G, then $(\chi, \psi) = |G|^{-1} \sum_{g \in G} \chi(g) \cdot \psi(g^{-1})$.

If H is a subgroup of G, then $\chi|H$ is the restriction of χ to H.

If θ is a class function on H, then θ^G is the class function on G induced by θ.

$\langle a, b, \cdots \rangle$ is the group generated by $a, b, \cdots$.

REFERENCES

[1] A. A. Albert, "Structure of Algebras," Amer. Math. Soc., Providence, R. I., 1961.

[2] M. Benard, Quaternion constituents of group algebras, Proc. Amer. Math. Soc. 30 (1971), 217-219.

[3] ________, On the Schur indices of characters of the exceptional Weyl groups, Ann. of Math. 94 (1971), 89-107.

[4] ________, The Schur subgroup, I, J. Algebra 22 (1972), 374-377.

[5] ________ and M. M. Schacher, The Schur subgroup, II, J. Algebra 22 (1972), 378-385.

[6] S. D. Berman, Representations of finite groups over an arbitrary field and over rings of integers, Izv. Akad. Nauk SSSR Ser. Mat. 30 (1966), 69-132.

[7] H. Boerner, "Darstellungen von Gruppen," Springer, Berlin, 1955.

[8] R. Brauer, Über Zusammenhänge zwischen arithmetischen und invariantentheoretischen Eigenschaften von Gruppen linearer Substitutionen, Sitzber. Preuss. Akad. Wiss. (1926), 410-416.

[9] ________, On the representation of a group of order g in the field of the g-th roots of unity, Amer. J. Math. 67 (1945), 461-471.

[10] ________, On Artin's L-series with general group characters, Ann. of Math. 48 (1947), 502-514.

[11] ________, Applications of induced characters, Amer. J. Math. 69 (1947), 709-716.

[12] ________, Representations of groups of finite order, Proc. Int. Cong. Math., vol. II, (1950),33-36.

[13] ________, On the algebraic structure of group rings, J. Math. Soc. Japan 3 (1951), 237-251.

[14] ________, Representations of finite groups, Lectures on Modern Mathematics, ed. T. L. Saaty, John Wiley & Sons, New York, 1963.

[15] C. W. Curtis and I. Reiner, "Representation Theory of Finite Groups and Associative Algebras, Wiley-Interscience, New York, 1962.

[16] M. Deuring, "Algebren," Springer, Berlin, 1935.

[17] B. Fein, Note on the Brauer-Speiser theorem, Proc. Amer. Math. Soc. 25 (1970), 620-621.

[18] ______, Realizability of representations in cyclotomic fields, Proc. Amer. Math. Soc. 38 (1973), 40-42.

[19] ______, Schur indices and sums of squares, (to appear).

[20] ______ and T. Yamada, The Schur index and the order and exponent of a finite group, J. Algebra 28 (1974).

[21] W. Feit, "Characters of Finite Groups," Benjamin, New York, 1967.

[22] K. L. Fields, On the Brauer-Speiser theorem, Bull. Amer. Math. Soc. 77 (1971), 223.

[23] __________, Two remarks on the group algebra of a finite group, Proc. Amer. Math. Soc. 30 (1971), 247-248.

[24] __________ and I. N. Herstein, On the Schur subgroup of the Brauer group, J. Algebra 20 (1972), 70-71.

[25] P. Fong, A note on splitting fields of representations of finite groups, Illinois J. Math. 7 (1963), 515-520.

[26] C. Ford, Some results on the Schur index of a representation of a finite group, Canad. J. Math. 22 (1970), 626-640.

[27] ______ and G. J. Janusz, Examples in the theory of the Schur group, (to appear).

[28] D. M. Goldschmidt and I. M. Isaacs, Schur indices in finite groups, (to appear).

[29] H. Hasse, "Zahlentheorie," 2nd edition, Akademie-Verlag, Berlin, 1963.

[30] B. Huppert, "Endliche Gruppen, I," Springer, Berlin, 1967.

[31] G. J. Janusz, The Schur index and roots of unity, Proc. Amer. Math. Soc. 35 (1972), 387-388.

[32] __________, The Schur group of cyclotomic fields, J. Number Theory, (to appear).

[33] P. Roquette, Realisierung von Darstellungen endlicher nilpotenter Gruppen, Archiv Math. 9 (1958), 241-250.

[34] I. Schur, Arithmetische Untersuchungen über endliche Gruppen libeare Substitutionen, Sitzber. Preuss. Akad. Wiss. (1906), 164-184.

[35] J.-P. Serre, "Corps locaux," 2nd ed., Hermann, Paris, 1968.

[36] L. Solomon, The representation of finite groups in algebraic number fields, J. Math. Soc. Japan 13 (1961), 144-164.

[37] ________, On Schur's index and the solutions of $G^n = 1$ in a finite group, Math. Zeit. 78 (1962), 122-125.

[38] W. Specht, Darstellungstheorie der alternierenden Gruppe, Math. Zeit. 43 (1938), 553-572.

[39] A. Speiser, Zahlentheoretische Sätze aus der Gruppentheorie, Math. Zeit. 5 (1919), 1-6.

[40] B. L. van der Waerden, "Algebra, II," 3rd edition, Springer, Berlin, 1955.

[41] T. Yamada, On the group algebras of metacyclic groups over algebraic number fields, J. Fac. Sci. Univ. Tokyo 15 (1968), 179-199.

[42] ________, On the group algebras of metabelian groups over algebraic number fields, I, Osaka J. Math. 6 (1969), 211-228.

[43] ________, On the group algebras of metabelian groups over algebraic number fields, II, J. Fac. Sci. Univ. Tokyo 16 (1969), 83-90.

[44] ________, On the Schur index of a monomial representation, Proc. Japan Acad. 45 (1969), 522-525.

[45] ________, Characterization of the simple components of the group algebras over the p-adic number field, J. Math. Soc. Japan 23 (1971), 295-310.

[46] ________, Central simple algebras over totally real fields which appear in Q[G], J. Algebra 23 (1972), 382-403.

[47] ________, Simple components of group algebras Q[G] central over real quadratic fields, J. Number Theory 5 (1973), 179-190.

[48] ________, The Schur subgroup of the Brauer group, I, J. Algebra 27 (1973), 579-589.

[49] ________, The Schur subgroup of the Brauer group, II, J. Algebra 27 (1973), 590-603.

[50] ________, Cyclotomic algebras over a 2-adic field, Proc. Japan Acad. 49 (1973), 438-442.

[51] ________, The Schur subgroup of a 2-adic field, J. Math. Soc. Japan 26 (1974), 168-179.

[52] ________, On a splitting field of representations of a finite group, Pacific J. Math., (to appear).

[53] ________, The Schur subgroup of a real quadratic field, I, Symposia Mathematica, (Proceedings of the conference on "Structure of algebraic fields" held at INDAM, Rome, April 5-10, 1973), Academic Press, London, (to appear).

[54] ________, The Schur subgroup of a p-adic field, J. Algebra, (to appear).

[55] ________, The Schur subgroup of a real quadratic field, II, (to appear).

[56] E. Witt, Die algebraische Struktur des Gruppenringes einer endlichen Gruppe über einem Zahlkörper, J. reine angew. Math. 190 (1952), 231-245.

[57] H. J. Zassenhaus, "The Theory of Groups," 2nd ed., Chelsea, New York, 1958.

Vol. 215: P. Antonelli, D. Burghelea and P. J. Kahn, The Concordance-Homotopy Groups of Geometric Automorphism Groups. X, 140 pages. 1971. DM 16,-

Vol. 216: H. Maaß, Siegel's Modular Forms and Dirichlet Series. VII, 328 pages. 1971. DM 20,-

Vol. 217: T. J. Jech, Lectures in Set Theory with Particular Emphasis on the Method of Forcing. V, 137 pages. 1971. DM 16,-

Vol. 218: C. P. Schnorr, Zufälligkeit und Wahrscheinlichkeit. IV, 212 Seiten. 1971. DM 20,-

Vol. 219: N. L. Alling and N. Greenleaf, Foundations of the Theory of Klein Surfaces. IX, 117 pages. 1971. DM 16,-

Vol. 220: W. A. Coppel, Disconjugacy. V, 148 pages. 1971. DM 16,-

Vol. 221: P. Gabriel und F. Ulmer, Lokal präsentierbare Kategorien. V, 200 Seiten. 1971. DM 18,-

Vol. 222: C. Meghea, Compactification des Espaces Harmoniques. III, 108 pages. 1971. DM 16,-

Vol. 223: U. Felgner, Models of ZF-Set Theory. VI, 173 pages. 1971. DM 16,-

Vol. 224: Revètements Etales et Groupe Fondamental. (SGA 1). Dirigé par A. Grothendieck XXII, 447 pages. 1971. DM 30,-

Vol. 225: Théorie des Intersections et Théorème de Riemann-Roch. (SGA 6). Dirigé par P. Berthelot, A. Grothendieck et L. Illusie. XII, 700 pages. 1971. DM 40,-

Vol. 226: Seminar on Potential Theory, II. Edited by H. Bauer. IV, 170 pages. 1971. DM 18,-

Vol. 227: H. L. Montgomery, Topics in Multiplicative Number Theory. IX, 178 pages. 1971. DM 18,-

Vol. 228: Conference on Applications of Numerical Analysis. Edited by J. Ll. Morris. X, 358 pages. 1971. DM 26,-

Vol. 229: J. Väisälä, Lectures on n-Dimensional Quasiconformal Mappings. XIV, 144 pages. 1971. DM 16,-

Vol. 230: L. Waelbroeck, Topological Vector Spaces and Algebras. VII, 158 pages. 1971. DM 16,-

Vol. 231: H. Reiter, L^1-Algebras and Segal Algebras. XI, 113 pages. 1971. DM 16,-

Vol. 232: T. H. Ganelius, Tauberian Remainder Theorems. VI, 75 pages. 1971. DM 16,-

Vol. 233: C. P. Tsokos and W. J. Padgett. Random Integral Equations with Applications to stochastic Systems. VII, 174 pages. 1971. DM 18,-

Vol. 234: A. Andreotti and W. Stoll. Analytic and Algebraic Dependence of Meromorphic Functions. III, 390 pages. 1971. DM 26,-

Vol. 235: Global Differentiable Dynamics. Edited by O. Hájek, A. J. Lohwater, and R. McCann. X, 140 pages. 1971. DM 16,-

Vol. 236: M. Barr, P. A. Grillet, and D. H. van Osdol. Exact Categories and Categories of Sheaves. VII, 239 pages. 1971. DM 20,-

Vol. 237: B. Stenström, Rings and Modules of Quotients. VII, 136 pages. 1971. DM 16,-

Vol. 238: Der kanonische Modul eines Cohen-Macaulay-Rings. Herausgegeben von Jürgen Herzog und Ernst Kunz. VI, 103 Seiten. 1971. DM 16,-

Vol. 239: L. Illusie, Complexe Cotangent et Déformations I. XV, 355 pages. 1971. DM 26,-

Vol. 240: A. Kerber, Representations of Permutation Groups I. VII, 192 pages. 1971. DM 18,-

Vol. 241: S. Kaneyuki, Homogeneous Bounded Domains and Siegel Domains. V, 89 pages. 1971. DM 16,-

Vol. 242: R. R. Coifman et G. Weiss, Analyse Harmonique Non-Commutative sur Certains Espaces. V, 160 pages. 1971. DM 16,-

Vol. 243: Japan-United States Seminar on Ordinary Differential and Functional Equations. Edited by M. Urabe. VIII, 332 pages. 1971. DM 26,-

Vol. 244: Séminaire Bourbaki - vol. 1970/71. Exposés 382-399. IV, 356 pages. 1971. DM 26,-

Vol. 245: D. E. Cohen, Groups of Cohomological Dimension One. V, 99 pages. 1972. DM 16,-

Vol. 246: Lectures on Rings and Modules. Tulane University Ring and Operator Theory Year, 1970-1971. Volume I. X, 661 pages. 1972. DM 40,-

Vol. 247: Lectures on Operator Algebras. Tulane University Ring and Operator Theory Year, 1970-1971. Volume II. XI, 786 pages. 1972. DM 40,-

Vol. 248: Lectures on the Applications of Sheaves to Ring Theory. Tulane University Ring and Operator Theory Year, 1970-1971. Volume III. VIII, 315 pages. 1971. DM 26,-

Vol. 249: Symposium on Algebraic Topology. Edited by P. J. Hilton. VII, 111 pages. 1971. DM 16,-

Vol. 250: B. Jónsson, Topics in Universal Algebra. VI, 220 pages. 1972. DM 20,-

Vol. 251: The Theory of Arithmetic Functions. Edited by A. A. Gioia and D. L. Goldsmith VI, 287 pages. 1972. DM 24,-

Vol. 252: D. A. Stone, Stratified Polyhedra. IX, 193 pages. 1972. DM 18,-

Vol. 253: V. Komkov, Optimal Control Theory for the Damping of Vibrations of Simple Elastic Systems. V, 240 pages. 1972. DM 20,-

Vol. 254: C. U. Jensen, Les Foncteurs Dérivés de $\varprojlim$ et leurs Applications en Théorie des Modules. V, 103 pages. 1972. DM 16,-

Vol. 255: Conference in Mathematical Logic - London '70. Edited by W. Hodges. VIII, 351 pages. 1972. DM 26,-

Vol. 256: C. A. Berenstein and M. A. Dostal, Analytically Uniform Spaces and their Applications to Convolution Equations. VII, 130 pages. 1972. DM 16,-

Vol. 257: R. B. Holmes, A Course on Optimization and Best Approximation. VIII, 233 pages. 1972. DM 20,-

Vol. 258: Séminaire de Probabilités VI. Edited by P. A. Meyer. VI, 253 pages. 1972. DM 22,-

Vol. 259: N. Moulis, Structures de Fredholm sur les Variétés Hilbertiennes. V, 123 pages. 1972. DM 16,-

Vol. 260: R. Godement and H. Jacquet, Zeta Functions of Simple Algebras. IX, 188 pages. 1972. DM 18,-

Vol. 261: A. Guichardet, Symmetric Hilbert Spaces and Related Topics. V, 197 pages. 1972. DM 18,-

Vol. 262: H. G. Zimmer, Computational Problems, Methods, and Results in Algebraic Number Theory. V, 103 pages. 1972. DM 16,-

Vol. 263: T. Parthasarathy, Selection Theorems and their Applications. VII, 101 pages. 1972. DM 16,-

Vol. 264: W. Messing, The Crystals Associated to Barsotti-Tate Groups: With Applications to Abelian Schemes. III, 190 pages. 1972. DM 18,-

Vol. 265: N. Saavedra Rivano, Catégories Tannakiennes. II, 418 pages. 1972. DM 26,-

Vol. 266: Conference on Harmonic Analysis. Edited by D. Gulick and R. L. Lipsman. VI, 323 pages. 1972. DM 24,-

Vol. 267: Numerische Lösung nichtlinearer partieller Differential- und Integro-Differentialgleichungen. Herausgegeben von R. Ansorge und W. Törnig. VI, 339 Seiten. 1972. DM 26,-

Vol. 268: C. G. Simader, On Dirichlet's Boundary Value Problem. IV, 238 pages. 1972. DM 20,-

Vol. 269: Théorie des Topos et Cohomologie Etale des Schémas. (SGA 4). Dirigé par M. Artin, A. Grothendieck et J. L. Verdier. XIX, 525 pages. 1972. DM 50,-

Vol. 270: Théorie des Topos et Cohomologie Etale des Schémas. Tome 2. (SGA 4). Dirigé par M. Artin, A. Grothendieck et J. L. Verdier. V, 418 pages. 1972. DM 50,-

Vol. 271: J. P. May, The Geometry of Iterated Loop Spaces. IX, 175 pages. 1972. DM 18,-

Vol. 272: K. R. Parthasarathy and K. Schmidt, Positive Definite Kernels, Continuous Tensor Products, and Central Limit Theorems of Probability Theory. VI, 107 pages. 1972. DM 16,-

Vol. 273: U. Seip, Kompakt erzeugte Vektorräume und Analysis. IX, 119 Seiten. 1972. DM 16,-

Vol. 274: Toposes, Algebraic Geometry and Logic. Edited by. F. W. Lawvere. VI, 189 pages. 1972. DM 18,-

Vol. 275: Séminaire Pierre Lelong (Analyse) Année 1970-1971. VI, 181 pages. 1972. DM 18,-

Vol. 276: A. Borel, Représentations de Groupes Localement Compacts. V, 98 pages. 1972. DM 16,-

Vol. 277: Séminaire Banach. Edité par C. Houzel. VII, 229 pages. 1972. DM 20,-

Vol. 278: H. Jacquet, Automorphic Forms on GL(2). Part II. XIII, 142 pages. 1972. DM 16,-

Vol. 279: R. Bott, S. Gitler and I. M. James, Lectures on Algebraic and Differential Topology. V, 174 pages. 1972. DM 18,-

Vol. 280: Conference on the Theory of Ordinary and Partial Differential Equations. Edited by W. N. Everitt and B. D. Sleeman. XV, 367 pages. 1972. DM 26,-

Vol. 281: Coherence in Categories. Edited by S. Mac Lane. VII, 235 pages. 1972. DM 20,-

Vol. 282: W. Klingenberg und P. Flaschel, Riemannsche Hilbertmannigfaltigkeiten. Periodische Geodätische. VII, 211 Seiten. 1972. DM 20,-

Vol. 283: L. Illusie, Complexe Cotangent et Déformations II. VII, 304 pages. 1972. DM 24,-

Vol. 284: P. A. Meyer, Martingales and Stochastic Integrals I. VI, 89 pages. 1972. DM 16,-

Vol. 285: P. de la Harpe, Classical Banach-Lie Algebras and Banach-Lie Groups of Operators in Hilbert Space. III, 160 pages. 1972. DM 16,-

Vol. 286: S. Murakami, On Automorphisms of Siegel Domains. V, 95 pages. 1972. DM 16,-

Vol. 287: Hyperfunctions and Pseudo-Differential Equations. Edited by H. Komatsu. VII, 529 pages. 1973. DM 36,-

Vol. 288: Groupes de Monodromie en Géométrie Algébrique. (SGA 7 I). Dirigé par A. Grothendieck. IX, 523 pages. 1972. DM 50,-

Vol. 289: B. Fuglede, Finely Harmonic Functions. III, 188. 1972. DM 18,-

Vol. 290: D. B. Zagier, Equivariant Pontrjagin Classes and Applications to Orbit Spaces. IX, 130 pages. 1972. DM 16,-

Vol. 291: P. Orlik, Seifert Manifolds. VIII, 155 pages. 1972. DM 16,-

Vol. 292: W. D. Wallis, A. P. Street and J. S. Wallis, Combinatorics: Room Squares, Sum-Free Sets, Hadamard Matrices. V, 508 pages. 1972. DM 50,-

Vol. 293: R. A. DeVore, The Approximation of Continuous Functions by Positive Linear Operators. VIII, 289 pages. 1972. DM 24,-

Vol. 294: Stability of Stochastic Dynamical Systems. Edited by R. F. Curtain. IX, 332 pages. 1972. DM 26,-

Vol. 295: C. Dellacherie, Ensembles Analytiques, Capacités, Mesures de Hausdorff. XII, 123 pages. 1972. DM 16,-

Vol. 296: Probability and Information Theory II. Edited by M. Behara, K. Krickeberg and J. Wolfowitz. V, 223 pages. 1973. DM 20,-

Vol. 297: J. Garnett, Analytic Capacity and Measure. IV, 138 pages. 1972. DM 16,-

Vol. 298: Proceedings of the Second Conference on Compact Transformation Groups. Part 1. XIII, 453 pages. 1972. DM 32,-

Vol. 299: Proceedings of the Second Conference on Compact Transformation Groups. Part 2. XIV, 327 pages. 1972. DM 26,-

Vol. 300: P. Eymard, Moyennes Invariantes et Représentations Unitaires. II. 113 pages. 1972. DM 16,-

Vol. 301: F. Pittnauer, Vorlesungen über asymptotische Reihen. VI, 186 Seiten. 1972. DM 18,-

Vol. 302: M. Demazure, Lectures on p-Divisible Groups. V, 98 pages. 1972. DM 16,-

Vol. 303: Graph Theory and Applications. Edited by Y. Alavi, D. R. Lick and A. T. White. IX, 329 pages. 1972. DM 26,-

Vol. 304: A. K. Bousfield and D. M. Kan, Homotopy Limits, Completions and Localizations. V, 348 pages. 1972. DM 26,-

Vol. 305: Théorie des Topos et Cohomologie Etale des Schémas. Tome 3. (SGA 4). Dirigé par M. Artin, A. Grothendieck et J. L. Verdier. VI, 640 pages. 1973. DM 50,-

Vol. 306: H. Luckhardt, Extensional Gödel Functional Interpretation. VI, 161 pages. 1973. DM 18,-

Vol. 307: J. L. Bretagnolle, S. D. Chatterji et P.-A. Meyer, Ecole d'été de Probabilités: Processus Stochastiques. VI, 198 pages. 1973. DM 20,-

Vol. 308: D. Knutson, λ-Rings and the Representation Theory of the Symmetric Group. IV, 203 pages. 1973. DM 20,-

Vol. 309: D. H. Sattinger, Topics in Stability and Bifurcation Theory. VI, 190 pages. 1973. DM 18,-

Vol. 310: B. Iversen, Generic Local Structure of the Morphisms in Commutative Algebra. IV, 108 pages. 1973. DM 16,-

Vol. 311: Conference on Commutative Algebra. Edited by J. W. Brewer and E. A. Rutter. VII, 251 pages. 1973. DM 22,-

Vol. 312: Symposium on Ordinary Differential Equations. Edited by W. A. Harris, Jr. and Y. Sibuya. VIII, 204 pages. 1973. DM 22,-

Vol. 313: K. Jörgens and J. Weidmann, Spectral Properties of Hamiltonian Operators. III, 140 pages. 1973. DM 16,-

Vol. 314: M. Deuring, Lectures on the Theory of Algebraic Functions of One Variable. VI, 151 pages. 1973. DM 16,-

Vol. 315: K. Bichteler, Integration Theory (with Special Attention to Vector Measures). VI, 357 pages. 1973. DM 26,-

Vol. 316: Symposium on Non-Well-Posed Problems and Logarithmic Convexity. Edited by R. J. Knops. V, 176 pages. 1973. DM 18,-

Vol. 317: Séminaire Bourbaki – vol. 1971/72. Exposés 400–417. IV, 361 pages. 1973. DM 26,-

Vol. 318: Recent Advances in Topological Dynamics. Edited by A. Beck, VIII, 285 pages. 1973. DM 24,-

Vol. 319: Conference on Group Theory. Edited by R. W. Gatterdam and K. W. Weston. V, 188 pages. 1973. DM 18,-

Vol. 320: Modular Functions of One Variable I. Edited by W. Kuyk. V, 195 pages. 1973. DM 18,-

Vol. 321: Séminaire de Probabilités VII. Edité par P. A. Meyer. VI, 322 pages. 1973. DM 26,-

Vol. 322: Nonlinear Problems in the Physical Sciences and Biology. Edited by I. Stakgold, D. D. Joseph and D. H. Sattinger. VIII, 357 pages. 1973. DM 26,-

Vol. 323: J. L. Lions, Perturbations Singulières dans les Problèmes aux Limites et en Contrôle Optimal. XII, 645 pages. 1973. DM 42,-

Vol. 324: K. Kreith, Oscillation Theory. VI, 109 pages. 1973. DM 16,-

Vol. 325: Ch.-Ch. Chou, La Transformation de Fourier Complexe et L'Equation de Convolution. IX, 137 pages. 1973. DM 16,-

Vol. 326: A. Robert, Elliptic Curves. VIII, 264 pages. 1973. DM 22,-

Vol. 327: E. Matlis, 1-Dimensional Cohen-Macaulay Rings. XII, 157 pages. 1973. DM 18,-

Vol. 328: J. R. Büchi and D. Siefkes, The Monadic Second Order Theory of All Countable Ordinals. VI, 217 pages. 1973. DM 20,-

Vol. 329: W. Trebels, Multipliers for (C, α)-Bounded Fourier Expansions in Banach Spaces and Approximation Theory. VII, 103 pages. 1973. DM 16,-

Vol. 330: Proceedings of the Second Japan-USSR Symposium on Probability Theory. Edited by G. Maruyama and Yu. V. Prokhorov. VI, 550 pages. 1973. DM 36,-

Vol. 331: Summer School on Topological Vector Spaces. Edited by L. Waelbroeck. VI, 226 pages. 1973. DM 20,-

Vol. 332: Séminaire Pierre Lelong (Analyse) Année 1971-1972. V, 131 pages. 1973. DM 16,-

Vol. 333: Numerische, insbesondere approximationstheoretische Behandlung von Funktionalgleichungen. Herausgegeben von R. Ansorge und W. Törnig. VI, 296 Seiten. 1973. DM 24,-

Vol. 334: F. Schweiger, The Metrical Theory of Jacobi-Perron Algorithm. V, 111 pages. 1973. DM 16,-

Vol. 335: H. Huck, R. Roitzsch, U. Simon, W. Vortisch, R. Walden, B. Wegner und W. Wendland, Beweismethoden der Differentialgeometrie im Großen. IX, 159 Seiten. 1973. DM 18,-

Vol. 336: L'Analyse Harmonique dans le Domaine Complexe. Edité par E. J. Akutowicz. VIII, 169 pages. 1973. DM 18,-

Vol. 337: Cambridge Summer School in Mathematical Logic. Edited by A. R. D. Mathias and H. Rogers. IX, 660 pages. 1973. DM 42,-

Vol: 338: J. Lindenstrauss and L. Tzafriri, Classical Banach Spaces. IX, 243 pages. 1973. DM 22,-

Vol. 339: G. Kempf, F. Knudsen, D. Mumford and B. Saint-Donat, Toroidal Embeddings I. VIII, 209 pages. 1973. DM 20,-

Vol. 340: Groupes de Monodromie en Géométrie Algébrique. (SGA 7 II). Par P. Deligne et N. Katz. X, 438 pages. 1973. DM 40,-

Vol. 341: Algebraic K-Theory I, Higher K-Theories. Edited by H. Bass. XV, 335 pages. 1973. DM 26,-

Vol. 342: Algebraic K-Theory II, "Classical" Algebraic K-Theory, and Connections with Arithmetic. Edited by H. Bass. XV, 527 pages. 1973. DM 36,-